AF445475

Nanoimprint Lithography Technology and Applications

Nanoimprint Lithography Technology and Applications

Editor

Michael Mühlberger

MDPI • Basel • Beijing • Wuhan • Barcelona • Belgrade • Manchester • Tokyo • Cluj • Tianjin

Editor
Michael Mühlberger
Functional Surfaces and
Nanostructures
PROFACTOR GmbH
Steyr
Austria

Editorial Office
MDPI
St. Alban-Anlage 66
4052 Basel, Switzerland

This is a reprint of articles from the Special Issue published online in the open access journal *Nanomaterials* (ISSN 2079-4991) (available at: www.mdpi.com/journal/nanomaterials/special_issues/nanoimprint_lithography).

For citation purposes, cite each article independently as indicated on the article page online and as indicated below:

LastName, A.A.; LastName, B.B.; LastName, C.C. Article Title. *Journal Name* **Year**, *Volume Number*, Page Range.

ISBN 978-3-0365-4482-3 (Hbk)
ISBN 978-3-0365-4481-6 (PDF)

Contents

About the Editor

Michael Mühlberger

Michael Mühlberger is Senior Scientist in the Functional Surfaces and Nanostructures department at PROFACTOR. He studied physics and received his PhD in semiconductor physics from the Johannes Kepler University Linz, Austria in 2003. He soon joined PROFACTOR, where he initiated activities related to nanoimprinting. From 2011 to 2013, he was the Head of the Functional Surfaces and Nanostructures department, and since 2013 he holds a Senior Scientist postition. Michael Mühlberger has coordinated several national and international research projects, among them the NILaustria research project cluster in the Austrian Nanoinitiative. His research interests cover the nanoimprinting process itself, as well as applications of nanoimprinting ranging from optics to life sciences.

Preface to "Nanoimprint Lithography Technology and Applications"

Nanoimprint Lithography (NIL) has been an interesting and growing field in recent years since its beginnings in the mid-1990s. Since that time, nanoimprinting has undergone significant changes and developments and nowadays is a technology used in R&D labs around the world as well as in industrial production processes. One of the exciting things about the nanoimprinting process is its remarkable versatility and broad range of applications. This reprint includes ten articles that represent a small glimpse of the challenges and possibilities of this technology.

Six contributions deal with nanoimprint processes aiming at specific applications, while the other four papers focus on more general aspects of nanoimprint processes or present novel materials. Several different types of nanoimprint processes are used: plate-to-plate, roll-to-plate, and roll-to-roll. Plate-to-plate NIL here also includes the use of soft and flexible stamps.

The impact of the geometry on the filling properties of the nanofeatures in UV-NIL is discussed in the contribution by Thanner and Eibelhuber, while Mayer and Scheer discuss in detail the influence of the initial layer's thickness on the nanoimprint process and pattern formation, for both thermal NIL and UV-NIL. Guiding charts to choose a convenient initial layer thickness and the theoretical background are presented. Aspects of stamp fabrication are addressed by Marumo et al., including a way to predict stamp lifetime. Müller et al. report a UV-curable polymer with surface-active thiol groups. This type of materials can open up additional applications and functionalities for nanoimprinting.

A novel concept of nanoimprint-fabricated microtiter plates for neuronal cells is presented by Lohse et al. Roll-to-roll nanoimprinting is used in this paper as a fabrication method. Taus et al. describe a fabrication process of master structures for the nanoimprint replication of biomimetic structures, which are inspired by the Morpho butterfly. How these masters are used is described in the next section of the book, where it is shown that nanoimprinting is capable of replicating complex undercut nanostructures. The work by Prajzler et al. presents the results on the roll-to-plate imprinting of optical waveguides for optical interconnect applications and more. Atthi et al. present the fabrication procedure and properties of superhydrophobic and oleophobic surfaces based on structures with a high aspect ratio and fabricated by roll-to-plate nanoimprinting. Haslinger et al. discuss organic thin-film detectors. A novel device concept is presented that makes use of a nanoimprinted substrate. The detector is sensitive with respect to the polarization and direction of the incident light.

In summary, the application fields in this reprint are broad and can be identified as plasmonics, superhydrophibicity, biomimetics, optics/datacom, and life sciences, showing the broad applicability of nanoimprinting. The sections on the nanoimprint process discuss filling and wetting aspects during nanoimprinting as well as materials for stamps and imprinting.

I hope that this compilation of papers is interesting and entertaining for the readers and that it will generate many new ideas.

Michael Mühlberger

Editor

 *nanomaterials*

Article

UV Nanoimprint Lithography: Geometrical Impact on Filling Properties of Nanoscale Patterns

Christine Thanner and Martin Eibelhuber *

EV Group, DI Erich Thallner Str. 1, 4782 St. Florian am Inn, Austria; c.thanner@evgroup.com
* Correspondence: m.eibelhuber@evgroup.com

Abstract: Ultraviolet (UV) Nanoimprint Lithography (NIL) is a replication method that is well known for its capability to address a wide range of pattern sizes and shapes. It has proven to be an efficient production method for patterning resist layers with features ranging from a few hundred micrometers and down to the nanometer range. Best results can be achieved if the fundamental behavior of the imprint resist and the pattern filling are considered by the equipment and process parameters. In particular, the material properties and pattern size and shape play a crucial role. For capillary force-driven filling behavior it is important to understand the influencing parameters and respective failure modes in order to optimize the processes for reliable full wafer manufacturing. In this work, the nanoimprint results obtained for different pattern geometries are compared with respect to pattern quality and residual layer thickness: The comprehensive overview of the relevant process parameters is helpful for setting up NIL processes for different nanostructures with minimum layer thickness.

Keywords: nanoimprint lithography; UV-NIL; SmartNIL

Citation: Thanner, C.; Eibelhuber, M. UV Nanoimprint Lithography: Geometrical Impact on Filling Properties of Nanoscale Patterns. *Nanomaterials* **2021**, *11*, 822. https://doi.org/10.3390/nano11030822

Academic Editor: Michael Mühlberger

Received: 11 February 2021
Accepted: 17 March 2021
Published: 23 March 2021

Publisher's Note: MDPI stays neutral with regard to jurisdictional claims in published maps and institutional affiliations.

1. Introduction

Since it was first mentioned in literature [1], nanoimprint lithography (NIL) emerged into an attractive patterning technique and developed considerably in terms of materials, process technology, and equipment. In the recent years, NIL technology has proven a wide range of capabilities, leading to its recommendation as a production suitable alternative lithography method. Particularly soft UV-NIL [2] is a very flexible and efficient technology applicable to a large variety of structure sizes and shapes [3], including complex shapes and 3D patterns [4] without compromising mass manufacturability.

Nowadays, UV-NIL is the method of choice for various emerging applications [5] due to the maturity of the technology (equipment, process, and materials) and its compatibility with semiconductor manufacturing environment. Even though NIL is still considered as niche technology compared to optical lithography, it already has a proven track record for volume production in fields like wafer level optics [6], augmented reality [7], and biomedical diagnostics [8]. In these areas, the key differentiator of this technology is the capability to pattern permanent functional layers and the ability to provide high-resolution patterns on large areas. These NIL benefits can overcome several limitations of other lithography techniques while being largely scalable in substrate size and production volume.

However, as a replication technique, it behaves fundamentally different compared to lithography techniques based on (UV) exposure. As depicted in Figure 1a,b, those are typically based on the principle that only certain areas are exposed either by using shadow masks or by direct writing. For these areas, the UV exposure triggers a chemical reaction within the resist, and unexposed areas are not changed. The pattern can then be easily revealed during a development step that, depending on the resist type, will remove the exposed resist from the exposed areas for positive resists or unexposed areas for negative resist [9]. In contrast, UV-NIL, as shown in Figure 1c, is typically using UV exposure of all resist on the substrate and only negative resists are used. Additionally, most (optical) lithography is performed contactless, with a certain proximity gap between the mask and the resist, whereas it is obviously inherent to NIL process to contact the surface and fill the structures of the stamp used for patterning. This means that, while for other techniques it is most important to control the pattern quality during exposure by the optical system, in contrast, for NIL it is crucial to control the filling process.

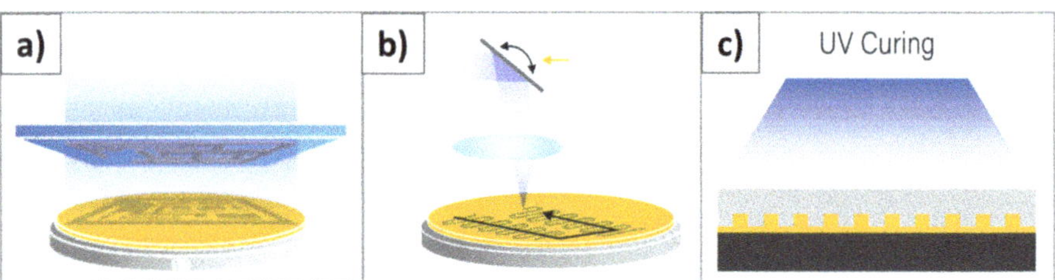

Figure 1. Schematic drawing of Ultraviolet (UV)-based lithographic techniques: (**a**) Optical lithography with a shadow mask, (**b**) direct writing, (**c**) UV-Nanoimprint Lithography (NIL).

The filling process has been extensively studied in previous work indicating to be a key contributor to certain failure modes as incomplete pattern replication, height, and residual layer variation [10,11]. Even though sometimes anticipated for spin-on UV-NIL, the results discussed by Sreenivasan and Cheng et al. refer to the high temperature and high-pressure regime of nanoimprint lithography by hot embossing. This type of filling is rather described as squeeze flow as elaborated by Rowland et al. [12]. The according defects in the presence of pattern and size variation for spin only thus have to be considered in this context as the shear resistance can cause various defects, including non-filled features and deformed features under pressure [10]. Other work such as that Song et al. [13,14] consider the filling in a low-pressure regime using a method referred as reverse NIL. In this case, the stamp is coated, and the patterns are transferred under pressure and heat afterwards. Consequently the filling and the pattern transfer are separated and not directly comparable to a to a conventual UV-NIL process.

For the UV-NIL, numerical descriptions as done by Taylor et al [15–18] and Yin et al. [19,20] take the next external forces as well as the capillary forces and the according surface energies into account, and thus seem to be the most applicable. However, some experimental findings like lateral redistribution of low viscous spin on resist and according changes in filling behavior are not fully explained. Particularly, defects occurring close to critical resist thickness for minimum residual layers cannot be fully explained. Previous work from Lee et al. [21] and Yasuda et al. [22] indicate that the filling was mainly investigated in a pressure-controlled regime rather than in the capillary regime. As will be shown in this work, state-of-the-art NIL stamp and resist combinations appear to have different surface energies and the filling is rather dominated by the fluidic properties.

Consequently, NIL processes have different prerequisites to achieve high quality patterning and it is essential to understand the influencing parameters and boundary conditions for reliable manufacturing. This work focuses on detailed analysis of the post-imprint residual layer, resist viscosity, and its flowing behavior (redistribution) as well as their respective impact on the filling behavior of various shapes of nanostructures.

2. Materials and Methods

The study is based on the EVG SmartNIL technology using an EVG7200 imprint system (EV Group, St. Florian am Inn, Austria). All imprints were done on full 8″ wafers, which were spin coated on an EVG101 coating system (EV Group, St. Florian am Inn, Austria). Two different resist materials have been used. EVG UV-NIL AS (EV Group, St. Florian am Inn, Austria) is used for the different geometric structures and was chosen for those tests as it shows typically very good filling behavior. Additionally, the layer thickness control can be easily achieved by dedicated versions optimized for different thickness at 2500 rpm and 60 s spin time as shown in Figure 2. For thicknesses below 50 nm, the solvent content has been manually adjusted to achieve a final thickness of only 20 nm.

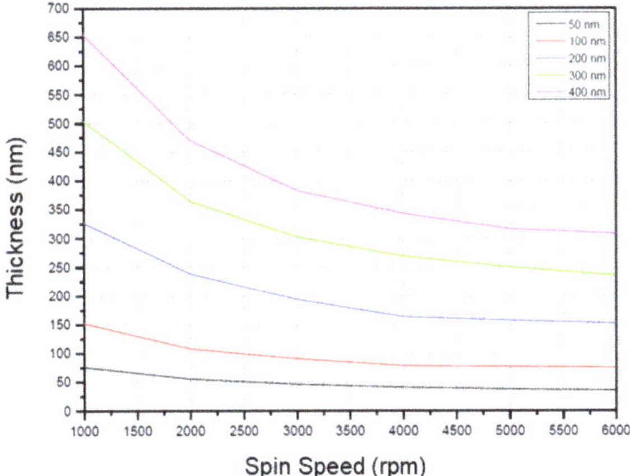

Figure 2. Spin curves of EVG UV-NIL AS with different solvent dilution. The solvent content is optimized for different thickness at 2500 rpm and 60 s spin time. The respective thickness values are shown in the inset.

The according spin speeds for the used thicknesses of 90 nm, 55 nm, and 20 nm were 2800 rpm, 1900 rpm, and 2000 rpm, respectively, for 1 min. All coatings were followed by a solvent release bake on a hot plate at 120 °C for 1 min.

For the second part of the study EVG UV-NIL E has been used as it was found that the viscosity could be also measured Tuning Fork Vibro Viscometer without additional solvent. The viscometer uses 2 mL samples to measure a viscosity range from 0.3–1000 cP (mPa·s). To have a solvent free reference is important for better understanding of the fluidic behavior as during the actual imprint all resists are considered solvent free due to the soft bake.

Both resists can be imprinted under comparable process conditions. Standard parameter at room temperature are 0 s delay time after full contact and exposure for 30 s at 300 mW/cm^2 from a 365 nm LED light source.

Additionally, specific UV-NIL working stamps have been used. The EVG NIL UV/AF working stamp material can be spin-coated and UV cured like other resins and does not need any additional treatment. The coating parameter were 3500 rpm for 5 min without any soft bake. To make the working stamp, the material was cured for 200 s with 300 mW/cm^2 at 365 nm.

For the characterization of the imprinted structures the Brooker Nanoscope AFM system has mainly been used. In detail, FASTSCAN C needles with 5 nm tip radius were used in Peak Force Tapping mode. Typical scan fields were 10×10 µm to 20×20 µm. For the investigation of larger areas brightfield images were made by a Zeiss Axiotron optical microscope with $5\times$ to $50\times$ magnification.

Additionally, to determine the viscosity of the EVG UV-NIL E resist, a Brookfield DV2TLV rotating viscometer was used. It allows to measure a viscosity range 15–30,000 mPa·s with a temperature sensing range of -100 °C to 300 °C. For the experiments, only a range from room temperature to 60 °C has been considered to determine the viscosity vs. the temperature.

3. Results

For the study of the UV-Nanoimprint process, the SmartNIL® technology has been used. This is a UV-NIL method performing a wafer level or full substrate imprint using transparent and flexible polymer working stamps. The complete wafer level process flow is shown in Figure 3. This includes the manufacturing of the working stamp (steps 1–4) and the actual imprint process (steps 5–8) [23,24]. As a starting point, for each imprinted pattern, there has to be a master stamp manufactured. The master stamp provides the actual topography of the desired structures for the replication and is most commonly, in particular for nanostructures, prepared on silicon using e-beam lithography and subsequent etching. However, for the imprint process, the actual patterning method and material are secondary if they are compatible with the materials used for replicating the working stamp. The replication process is using a working stamp, which is produced based on the master stamp. The working stamp fabrication starts with the coating of an anti-sticking layer (ASL) applied to the master stamp: This significantly increases the contact angle of the surface and minimizes the surface energy for the working stamp material, resulting in improved detaching of the working stamp after production. Secondly, the working stamp material is dispensed by spin coating and a flexible backplane is mounted. After contact with the master stamp the working stamp polymer material is cured and the negative of the master stamp surface topography is replicated in the working stamp polymer. Finally, the working stamp can be demolded after solidification, and is then used for the actual nanoimprint process.

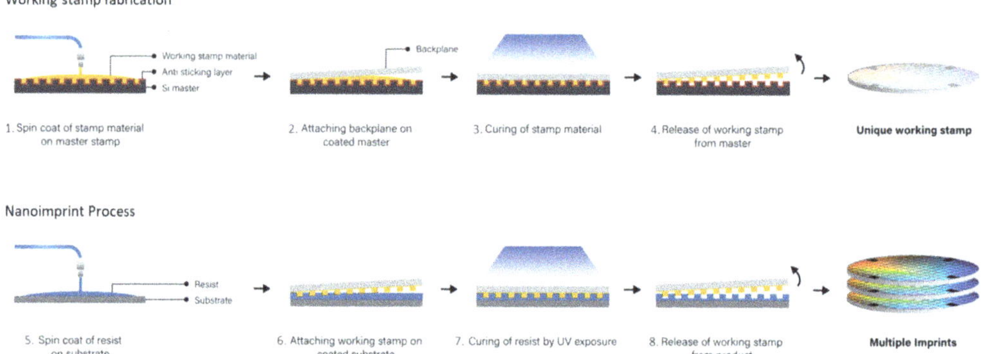

Figure 3. Schematic illustration of the SmartNIL process flow. Steps 1–4 show the typical manufacturing steps for creating a transparent working stamp. Steps 5–8 illustrate the actual imprint process which can be repeated multiple times.

For nanoimprinting and effectively patterning the target substrate, the nanoimprint resist is spin coated onto the respective surface and a soft bake is performed. The resist thickness and uniformity are crucial for the filling of the structures: This also has to consider the aspect ratio and fill factor of the master design. Subsequently, the flexible working stamp is fully conformal applied to the substrate, which is enabled by the flexible backplane support. While the working stamp and the wafer surface are in contact, the stamp structures are filled by capillary forces. The key parameters for this filling are further investigated in this paper: The amount of resist, the resist viscosity, and the flow behavior. Those parameters have significant impact on the residual layer thickness and the replication quality in terms of pattern fidelity and defects. Finally, the resist is cured under UV exposure and the working stamp can be demolded. Multiple imprints can be performed with a single working stamp as it can be reused.

Even though the above-discussed full SmartNIL imprint process is split in two parts of manufacturing the working stamp and subsequently imprinting the wafer, the entire process can be implemented on the same equipment. This work was performed on the EVG®101 spin coating and the EVG®7200 SmartNIL system using 200 mm wafers.

In order to study the filling properties based on the available imprint resist material, viscosity, and flow behavior, an e-beam written and etched silicon master was used. On this master, several arrays with different structure geometries as described in Figure 4a–e are present. The pattern investigated pattern sizes were 300 nm with a depth of approximately 110 nm have been replicated by SmartNIL and according SEM images are shown in Figure 4f–j.

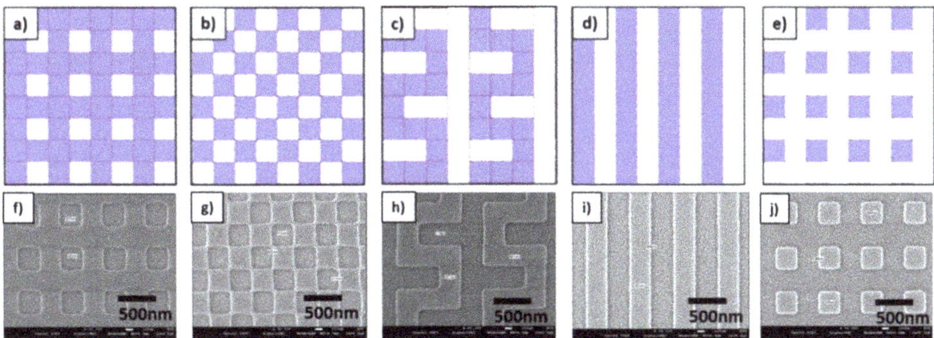

Figure 4. Schematic illustration of the different geometrical layouts used to investigate the filling behavior of nanostructures: (**a**) Crossbar, (**b**) checkerboard, (**c**) meanders, (**d**) line and space (L/S), and (**e**) pillars. Pattern dimensions of 300 nm with a depth of 110 nm have been selected for comparison and respective SEM images of SmartNIL replications are shown from (**f**–**j**).

For imprint processes it is very often crucial to control the layer uniformity and resist thickness in order to achieve the desired imprint quality. In particular, this is required if the residual layer has to be minimized in order to facilitate subsequent etching processes using the imprinted features as mask. Therefore, the impact of the resist thickness on the filling behavior and the resulting residual layer has been investigated in more detail.

At first, a crossbar structure was investigated with fill factor of 75% and all connected patterns. From the theoretical calculations according to formula $d_{resist} = ((pitch)^2 - (width)^2) \times depth/(pitch)^2$, the ideal resist thickness for minimum residual layer is considered 82.5 nm. The same value can be estimated by just using 75% of the pattern. Practically, as it is not possible to achieve zero residual layer with typical imprint resists, the actual resist thickness should be chosen at least a few nm above the ideal value. As shown in the Atomic Force Microscope (AFM) images in Figure 5, the imprint process was performed with different resist thickness: 90 nm, 85 nm, and 75 nm, respectively. In order to achieve the different layer thickness, a single resist type (EVG UV-NIL A) was used, but the spin coating parameters were varied. For 90 nm layer thickness was used a speed of 1500 rpm for 60 s, while for 85 nm and 75 nm, 1800 rpm and 2500 rpm were used, respectively.

In this way the amount of the resist was heuristically lowered to achieve a minimum residual layer and was further reduced to a level where the filling of the imprint pattern was limited by the available resist volume. Working intuitively is often wrong considering that this will rather impact the vertical filling of the structures. This is only true for macroscopic structures where the external applied printing force is domination the filling behavior but cannot be extrapolated to the nanoscale region. For these structures, capillary forces start to dominate the process [10] and change the filling behavior significantly. Therefore, Figure 5 shows imprints where the resist layer was set to be (a) slightly more, (b) approximately equal to, and (c) less than required filling volume. In detail it is observable in Figure 5a that for the 300 nm crossbar pattern a complete filling of the structures could be achieved with a

90 nm thick resist layer. When lowering the resist thickness to 85 nm as shown in Figure 5b, minor defects were occurring randomly. Those defects significantly increase when the resist layer thickness is further decreased to 75 nm. This shows that the nanostructures will be filled in some areas completely while in other areas the resist is completely missing. Thus, it is concluded that the resist volume will be redistributed and the capillary forces dominate the filling behavior.

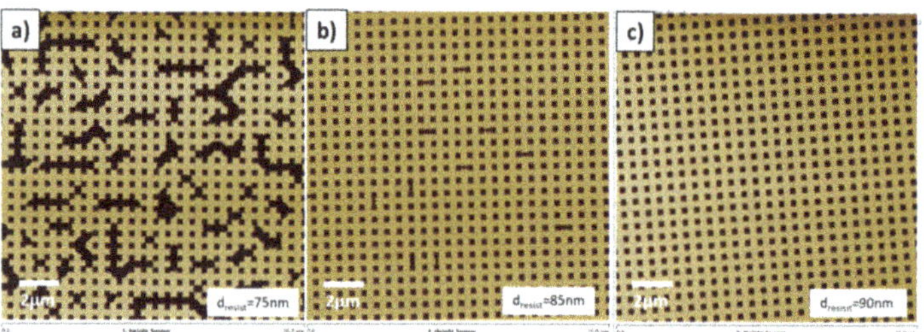

Figure 5. AFM images of the imprinted crossbar pattern using different resist layer thickness: (**a**) Fully replicated pattern with enough resist volume (d_{resist} = 90 nm), (**b**) Imprint wits slightly less resist volume showing minor defects (d_{resist} = 85 nm), and (**c**) Imprint with too less resist volume showing significant defects (d_{resist} = 75 nm).

Next to the filling of the stamp structure it is also important to understand the impact on the residual layer control. The residual layer thickness has been determined: part of the imprinted structure was removed by scratching down to substrate surface and the step created was investigated using AFM measurements. Figure 6a,b show the topography measured for the 90 nm and respectively 75 nm samples. The measurement is considering that no resist is left for the scratched areas and the thickness is measured from the substrate surface. The residual layer thickness is defined by the height difference to the valleys between the crossbar structures. In this way for the completely filled 90 nm structure an average residual layer thickness of about 5 nm was observed. In comparison the 75 nm sample the average residual layer thickness is about 3 nm. Interestingly, this seems also to be valid for areas with missing structures, where it could be assumed that all the resist was transferred to the filled areas. Even though this finding has to be verified in more detail with different imprint resist, it clearly shows the limitations of residual layer control in the <10 nm range and additionally confirms that with conventional materials and processes a residual layer free imprinting cannot be achieved.

Notably, also a second aspect indicated above could be approved by this measurement. It was assumed from the top view measurements above the full pattern height achieved for the 75 nm layer thickness by redistribution of the resist. As can be seen in the AFM topography data the pattern height of 110 nm is similar to the completely filled sample with 90 nm layer thickness. This proofs that the tendency to fill the topography rather completely is more pronounced than keeping a uniform layer beneath the pattern and fill them only partly in height as observed in the pressure regime [17,18]. Thus it is concluded the dominating nature for the structure filling of the nanostructures is given by the capillary regime.

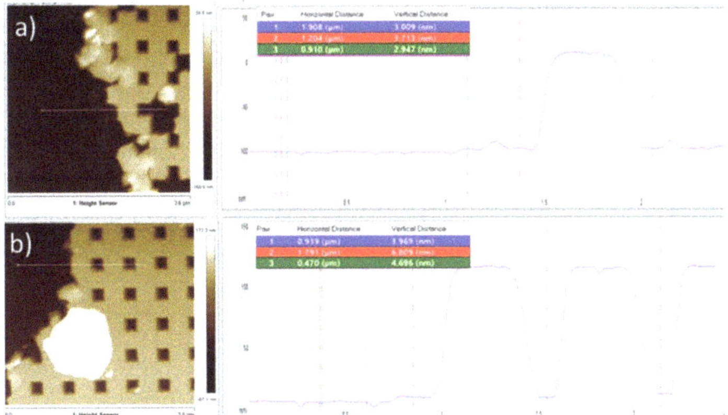

Figure 6. Atomic Force Microscope (AFM) top view image and height profile of the imprinted crossbar pattern used to measure the residual layer thickness for: (**a**) Complete filling (d_{resist} = 90 nm) and (**b**) incomplete filling (d_{resist} = 75 nm).

As the observed behavior might be linked to the actual structure design, the impact on different pattern shapes has been investigated as well (Figure 7). As the different structures do not have the same filling factor, the resist layer thickness has been modified for the different designs. The crossbar design already discussed in Figures 5 and 6 has a fill factor of 83% and the structures are all connected. When comparing Figure 7a of the fully replicated structure with Figure 7b of the partially filled structure it seems that the geometry also influences the defect pattern. When the amount of resist is lowered below the optimum these defects occur more or less at random positions but overall rather uniformly distributed. The defects seem to be mostly connected with the holes in the crossbar pattern and interacting with them but often with less defined roundish borders. Additionally, it appears that the non-filled area also rather tends to expand and connect, resulting in large non-patterned areas with sharp borders to fully patterned, non-defective areas.

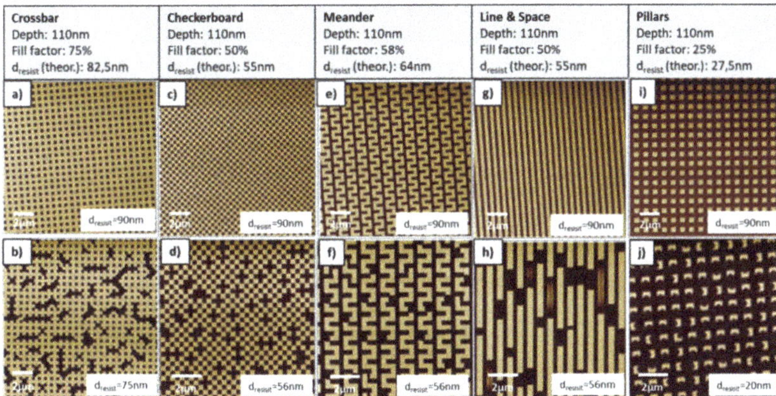

Figure 7. Top view images showing a comparison between different geometries after imprinting: (**a**,**b**) Crossbar, (**c**,**d**) checkerboard, (**e**,**f**) meanders, (**g**,**h**) line and space (L/S), and (**i**,**j**) pillars. Top row shows the fully replicated patterns using optimum resist volume. Bottom row shows the observed defects if less than optimum resist volume is used. The critical dimension for all patterns geometries was 300 nm.

In order to investigate the impact of other pattern geometries independently from the fill factor, three different designs with the same critical dimension of 300 nm and comparable fill factor have been analyzed. These are the checkerboard with a fill factor of 50%, the meander structure with a fill factor of 58%, and the L/S structure with 50% fill factor. For these structures, the resist thickness had to be lowered until incomplete filling appeared. For all three structures, a critical resists layer thickness of 56 nm was observed.

The checkerboard structure shown in Figure 7c,d is geometrically the most similar to the crossbar structure as it has similar holes, and the pattern area is still quasi connected through the edges of the pattern. As can be seen in Figure 7d, in this case, the defects are randomly but quite uniformly distributed. However, as the resist cannot freely flow, the defects are mainly complete missing patterns with well-defined borders, which appear to be more isolated.

Figure 7e,f show the impact on the meander structures: This geometry is still relatively close to the crossbar pattern but in this case, the valleys are also connected over a larger area. As it can be seen in Figure 7f, the defects are significantly different as compared to Figure 7d. While defects of the checkerboard in Figure 7d tended to be square-shaped, in contrast to the meander structure in Figure 7f, the tendency of the defects is clearly to be more rounded. This rounded shape is independent if the pattern starts at a corner or in the middle of a meander line and indicates that the defects start from a single point and have isotropic behavior if not influenced by the pattern. The heuristic comparison of the behavior of the checkerboard and the meander structures to the crossbar pattern shows a good overlap.

The change of the structure to L/S as shown in Figure 7g,h (50% filling factor) limits the redistribution of the resist in 1 dimension. As it can be seen also for this case, the incomplete filled structures show defects also only along a single direction. Specifically, it creates intersected lines with completely filled and unfilled parts. Interestingly, there are also some sections where the lines appear to be partly filled in height. Even though this seems not to be the dominating effect, it can be clearly seen that this intermediate filling exists as well. Even though not completely verified during this study, it is assumed that for these areas, a metastable state was fixed during curing. Further tests would be needed to verify if, with longer filling time or lower viscosity, these parts would have merged with already fully filled areas as well. Alternatively, it could also indicate a minimum line length to achieve complete filling.

Finally, the impact of too-thin resist layers was investigated for the inverse geometry of the crossbar structure in Figure 7i,j. The filling factor of this structure is only 25% and the according critical resist thickness is 27.5 nm. Additionally, there are no connections between the patterns that could support the redistribution of the resist. As for all other patterns studied in this series, it can be clearly seen that complete filling and a good pattern replication was also possible for the pillar structure with a 90 nm thick resist layer. To achieve the critical regime with clearly observable defects, the resist thickness was lowered to only 20 nm. As shown in Figure 7j, for these isolated structures it was not possible to redistribute the resist from one pattern to the other and no completely missing structures were observed. Consequently, to compensate for the missing resist volume every single pillar shows a small defect. This is quite remarkable, that even on this small area a resist redistribution similar as for the other patterns could be shown. Due to the capillary filling behavior, it is often experienced that small pillars rather get stuck in the stamp as there is no strong adhesion to the surface via the residual layer given anymore.

To understand if this effect is still present for larger pillars, a master with a structure of round pillars was replicated. The diameter of the pillars was 1 µm, with a height of 1.3 µm. The AFM images in Figure 8 show clearly a different behavior when compared to the 300 nm pillars with 110 nm depth. The AFM top view image in Figure 8a shows that for this pattern, no defects occur if the resist layer is too thin. However, as the volumes of the master and the offered resist must match, a different type of filling must occur and thus the structures will not be filled to the top. This fact is confirmed by the results shown in Figure 8b,c, where the structure height of about 620 nm is approximately half of the depth provided by the master template. All the pillars have about the same height and are approximately half filled. In addition, a distinct dimple in the center of every pillar is observed. This exactly fits to the expectations for capillary filling in case of macroscopic structures.

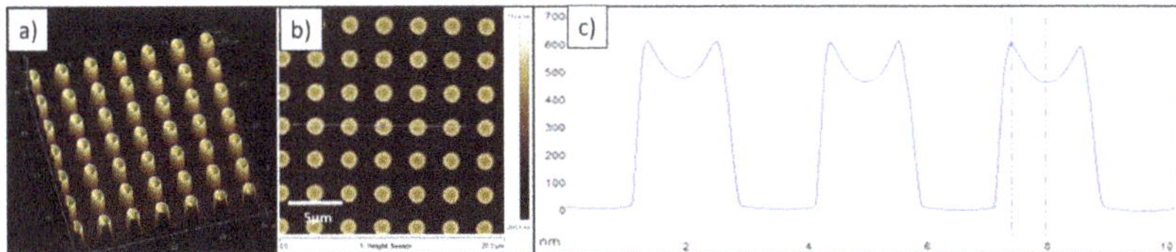

Figure 8. AFM images showing incomplete filling behavior of pillars with 1 µm diameter and 1.3 µm height. (**a**) Tilted view, (**b**) top view, and (**c**) height profile. In the top view, the line scan for the height profile can be seen.

As in the above study, a lateral redistribution of the resist was observed within the pattern area, it was also of interest to investigate impact on the surrounding area. Therefore, a master with 400 nm L/S pattern with fields separated by about 1 mm non-patterned area was chosen. Within the pattern area, there is a uniform 50% fill factor but at the border there is a change of the fill factor to 100%. It is considered that in this area, the lateral flow of the resist is not hindered by the pattern and thus shows approximately the free path length of the resist redistribution. As it can be observed in the optical microscope images in Figure 9a and according to the detail picture in Figure 9b, there is a large lateral redistribution of the resist. While the dark areas with the L/S pattern are homogenously filled, it can be clearly seen that at the border of these structures in the non-patterned bright area large defects occurred. In the detailed image in Figure 9b, it is clearly observable that there is a redistribution of the resist within a range of several tens of micrometers. The AFM image in Figure 9c confirms this redistribution while showing defect areas next to the completely filled L/S patterns. Interestingly, the shape of the defects is comparable to an elongated droplet like structure. Generally, the shape and the distance of the defects show some unexpected behavior. There are no defects directly at the border, but some are present in the border vicinity. However, the largest defects are just a few tens of micrometers away from the patterned area. This is yet unclear and might be intermixed with other effects not only based on the lateral redistribution of the resist to fill the patterned area. However, it could be proven that in non-patterned areas there is a long-range effect on the filling of the structures. This also implies that the lateral flow of the resist will also be for patterned areas will be on longer range until the required resist volume is consumed.

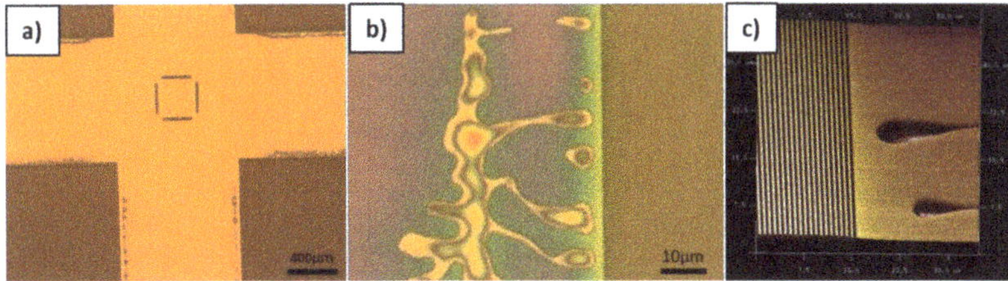

Figure 9. (**a**) Optical microscope image of imprinted L/S arrays (dark) and non-pattered areas (bright), (**b**) detail optical microscope image of defects occurring due to too low resist volume, and (**c**) AFM image of the imprinted L/S pattern showing the neighboring defects.

Obviously, it is not desired to have such defects in areas without patterns as well and thus it requires process optimizations to overcome this issue. As it is indicated that this is again strongly related to the fluidic properties, two key process parameters have been investigated on its effect on the border defects. As there are no structures where the defects occur it is obvious the capillary forces of the nanostructures are dominantly gaining material from this area. Even though there is enough material available in the unpatterned area, the defects are not compensated immediately if there is no delay time between full contact and UV exposure. For this process conditions, the defect area is well pronounced all around the edge zones of the patterns in a distinct distance from the L/S structure as depicted in Figure 10a. Adding delay time allows the material to redistribute better and to close the defects. It is easily observable in Figure 10a–d that the defects are vanishing more and more with increasing delay time from 30 s–90 s. At delay time of 90 s the defected zone is completely gone, showing that this effect is clearly related fluidic properties of the material. In Figure 10e–h it could be shown that by increasing the temperature to 30 °C the filling times could be reduced significantly. This is indicating the expected viscosity change of the resist at elevated temperatures. Thus, in a third test series, the viscosity of the resist was further lowered by increasing the temperature systematically without adding delay time. For all temperature dependent tests, a heated chuck was used, which can reach temperatures up to 60 °C. In Figure 10i,l,h, it is shown that increased temperature has a comparable influence as the delay time. The defects are strongly reduced by lowering the viscosity and fully vanished at 60 °C. This clearly proves that the resist redistribution is strongly liked to given filling and resist viscosity.

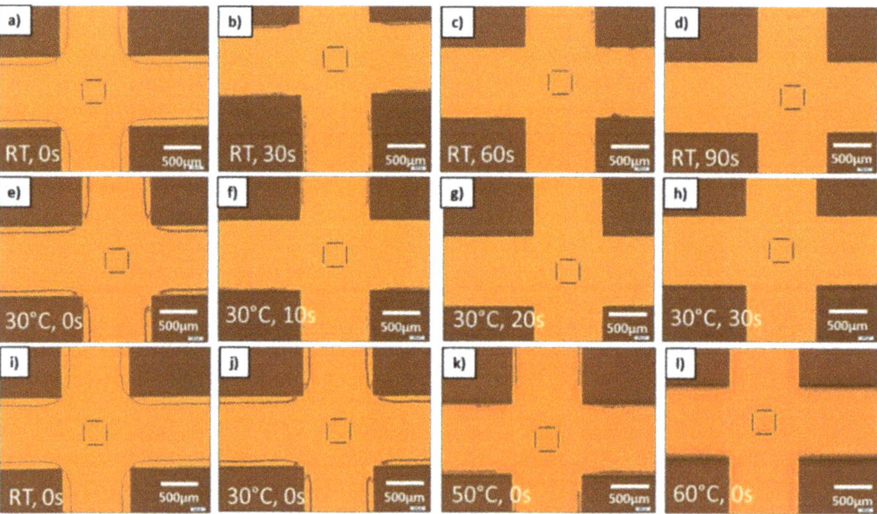

Figure 10. Filling behavior matrix for different temperatures and delay times until the defects completely. (**a–d**) Constant room temperature and increasing delay time in 30 s steps. (**e–h**) Constant elevated temperature at 30 °C and delay time increase by 10 s steps. (**i–l**) Constant delay time of 0 s and increasing temperature from RT up to 60 °C.

The dependence between delay time and imprint temperature is plotted in Figure 11 showing all series that have been investigated. The blue data series shows the time needed to have ideal imprint at a certain temperature. It is significantly dropping with increased temperature. This behavior correlates very well with the viscosity change measured within temperature range using a viscosimeter. Thus, for setting up NIL processes particularly with less known materials, temperature can be a crucial parameter to optimize the process and to enable proper replication and filling to the structures.

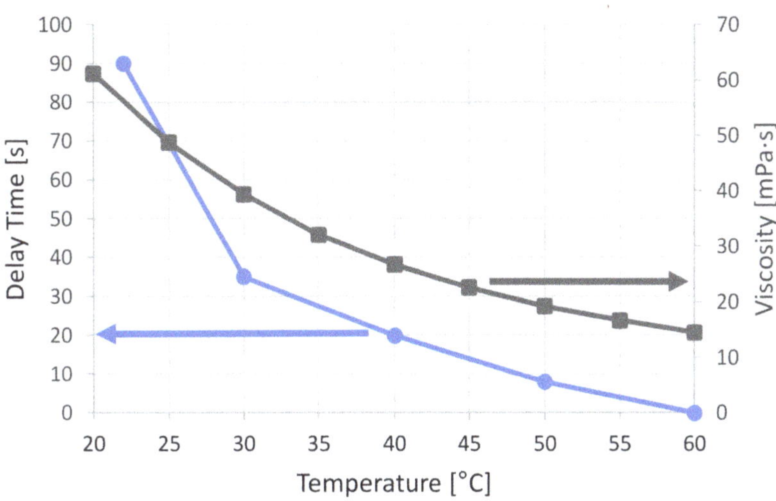

Figure 11. (**Left**) Plot of the required delay time versus the imprint temperature for defect free replication. (**Right**) Temperature dependence of the resist viscosity.

Overall, it could be demonstrated that the imprint process with state of the art material combinations are rather driven by capillary forces and not pressure driven as often found and discussed in earlier work [10–12] Thus, the fluidic properties, surface energies, and according capillary forces have a significant impact on the imprinting process and have to be taken more into account for the description of the UV-NIL process. Even though the capillary regime has been numerically described by Taylor et al. [13–16] lateral redistribution is mainly discussed in context of droplet spreading for layers dispensed by inkjet coating and laterally confined stamps used for step and repeat NIL [15,25]. According to the observations in this work, the lateral redistribution of full area UV-NIL imprints should be considered differently. It seems that those step and repeat approach is still best understood in the pressure regime and the main criteria for redistribution is the droplet spreading [15]. Wafer level processes like SmartNIL use flexible stamps and thus apply only comparably low force. Thus, state-of-the-art imprint materials are designed to work best in the capillary force dominated regime. Additionally, due to the much larger dimensions full area NIL takes more time until full area contact. This will allow certain redistribution also during the contact phase. Consequently, even without delay time between full contact and exposure this effect has been observed more easily. In particular, when aiming for a minimum residual layer, a critical resist thickness can be observed where lateral effects are strongly influencing the process control. An additional contrast is given by the fact that the spin coated areas can provide an additional resist reservoir for lateral redistribution at the boarders of the pattern areas. As described in this work, depending on the resist viscosity and the delay time, significant resist redistribution has to be considered on length scales of tens of μm. This shows the fluidic properties of the resist and the capillary forces are also crucial if there are major local changes of the patterns and according filling factors. If the viscosity or the given time are too low it is likely that defects with missing resist in those redistribution zones can occur. Therefore, improved process conditions with increased delay times or reduced viscosity can help to overcome this. Generally, it could be proven that low force techniques like SmartNIL, which support the inherent filling behavior of the resist, are very well suited to lower the layer thickness to critical condition and allow to achieve very thin layer imprints with high residual layer control. Temperature-controlled chucks can further improve the filling behavior to optimize the process conditions for high-quality imprints.

4. Conclusions

The study of minimizing the resist thickness even below the volume requirement for complete replication of the structures proves the filling of the structures is strongly dependent on fluidic properties of the resist and the capillary forces. This is of particular interest when the provided resist volume is at the limits for completely filling the structures. It is also clearly shown that at the nanoscale significant redistribution of the resist is present without external force. This general behavior is observed by filling studies showing comparable effects for all investigated structure geometries. The larger the connected areas, the more pronounced the lateral redistribution of the resist is. Connected structures thus tend to have more extended defects while isolated structures have a defect on every pattern. However, this behavior is only true if the geometrical dimensions are below a certain limit. For larger structures, there is no lateral redistribution and the structures get only shallower but still showing typical indications for capillary filling.

Additionally, it could be shown the capillary forces have a significant and long-range impact also in lateral direction. For areas with significant changes of the pattern density, edge effects occur showing evidence that resist is flowing from one to the other area by leaving defects in these zones. These flow defects can be reduced or completely avoided by prolonged delay times or reduced resist viscosity by using a heated chuck.

Summarizing, this has study shown that the fluidic properties of the resist and the pattern geometry have to be considered to set up reliable manufacturing processes. This is particularly to be considered for nanoscale structures with very thin residual layers, and the filling behavior can deviate significantly from expectations based on macroscopic experience.

Author Contributions: Conceptualization, C.T. and M.E.; Data curation, C.T.; Methodology, M.E.; Writing—original draft, M.E. All authors have read and agreed to the published version of the manuscript.

Funding: This research received no external funding.

Institutional Review Board Statement: Not applicable

Informed Consent Statement: Not applicable.

Data Availability Statement: Data is contained within the article or supplementary materials.

Acknowledgments: The authors would like to thank to IMS Chips for providing the master template with different pattern geometries which were used for this study.

Conflicts of Interest: The authors declare no conflict of interest.

References

1. Chou, S.Y.; Krauss, P.R.; Renstrom, P.J. Imprint Lithography with 25-Nanometer Resolution. *Science* **1996**, *272*, 85–87. [CrossRef]
2. Glinsner, T.; Plachetka, U.; Matthias, T.; Wimplinger, M.; Lindner, P. Soft UV-Based Nanoim-Print Lithography for Large-Area Imprinting Applications. In Proceedings of the SPIE, Emerging Lithographic Technologies XI, San Jose, CA, USA, 27 February–1 March 2007; Volume 6517, p. 651718.
3. Schift, H. Nanoimprint lithography: An old story in modern times? A review. *J. Vac. Sci. Technol. B Microelectron. Nanometer Struct.* **2008**, *26*, 458. [CrossRef]
4. Schift, H. Nanoimprint lithography: 2D or not 2D? A review. *Appl. Phys. A* **2015**, *121*, 415–435. [CrossRef]
5. Eibelhuber, M.; Uhrmann, T.; Glinsner, T.; Lindner, P. Nanoimprint Lithography enables cost effective photonics production. *Photonics Spectra* **2015**, *49*, 34–37.
6. Kast, M. High Precision Wafer level optics Fabrication and Integration. *Photonics Spectra* **2010**, *44*, 34–36.
7. Thanner, C.; Dudus, A.; Treiblmayr, D.; Berger, G.; Chouiki, M.; Martens, S.; Jurisch, M.; Hartbaum, J.; Eibelhuber, M. Nanoimprint lithography for augmented reality waveguide manufacturing. In *Proceedings of the Optical Architectures for Displays and Sensing in Augmented, Virtual and Mixed Reality (AR, VR, MR), San Francisco, CA, USA, 2–4 February 2020*; SPIE-International Society for Optics and Photonics: Bellingham, WA, USA, 2020; Volume 11310, p. 1131010.
8. Dielacher, B.; Eibelhuber, M.; Uhrmann, T. High-volume processes for next-generation biotechnology devices. *Solid State Technol.* **2016**, *59*, 11–16.
9. Levinson, H.J. *Principles of Lithography*; SPIE Press: Bellingham, WA, USA, 2001.

10. Sreenivasan, S. Nanoimprint lithography steppers for volume fabrication of leading-edge semiconductor integrated circuits. *Microsyst. Nanoeng.* **2017**, *3*, 17075. [CrossRef] [PubMed]
11. Cheng, X.; Guo, L.J. One-step lithography for various size patterns with a hybrid mask-mold. *Microelectron. Eng.* **2004**, *71*, 288–293. [CrossRef]
12. Rowland, H.D.; King, W.P.; Pethica, J.B.; Cross, G.L.W. Molecular Confinement Accelerates Deformation of Entangled Polymers during Squeeze Flow. *Science* **2008**, *322*, 720–724. [CrossRef] [PubMed]
13. Song, J.; Lu, H.; Foreman, K.; Li, S.; Tan, L.; Adenwalla, S.; Gruverman, A.; Ducharme, S. Ferroelectric polymer nanopillar arrays on flexible substrates by reverse nanoimprint lithography. *J. Mater. Chem. C* **2016**, *4*, 5914–5921. [CrossRef]
14. Song, J.; Lu, H.; Li, S.; Tan, L.; Gruverman, A.; Ducharme, S. Fabrication of ferroelectric polymer nanostructures on flexible substrates by soft-mold reverse nanoimprint lithography. *Nanotechnology* **2015**, *27*, 15302. [CrossRef] [PubMed]
15. Taylor, H.; Boning, D. Towards nanoimprint lithography-aware layout design checking. *SPIE Adv. Lithogr.* **2010**, *7641*, 76410. [CrossRef]
16. Taylor, H.; Wong, E. Fast Simulation of Nanoimprint Lithography: Modelling Capillary Pressures during Resist Deformation. In Proceedings of the 10th International Conf. Nanoimprint and Nanoprint Technology, Jeju, Korea, 19–21 October 2011.
17. Taylor, H.; Smistrup, K.; Boning, D. Modeling and simulation of stamp deflections in nanoimprint lithography: Exploiting backside grooves to enhance residual layer thickness uniformity. *Microelectron. Eng.* **2011**, *88*, 2154–2157. [CrossRef]
18. Taylor, H.K. Defectivity Prediction for Droplet-Dispensed UV Nanoimprint Lithography, Enabled by Fast Simulation of Resin Flow at Feature, Droplet, and Template Scales. In Proceedings of the Alternative Lithographic Technologies VIII, San Jose, CA, USA, 22–25 February 2016; Volume 9777, p. 97770E.
19. Yin, M.; Sun, H.; Wang, H. Resist Filling Study for UV Nanoimprint Lithography Using Stamps with Various Micro/Nano Ratios. *Micromachines* **2018**, *9*, 335. [CrossRef] [PubMed]
20. Yin, M.; Sun, H.; Wang, H. Effect of stamp design on residual layer thickness and contact pressure in UV nanoimprint lithog-raphy. *Micro Nano Lett.* **2018**, *13*, 887–891. [CrossRef]
21. Lee, H.; Ro, H.W.; Soles, C.L.; Jones, R.L.; Lin, E.K.; Wu, W.; Hines, D.R. Effect of imprinting pressure on residual layer thickness in ultraviolet nanoimprint lithography. *J. Vac. Sci. Technol. B* **2005**, *23*, 1102–1106. [CrossRef]
22. Yasuda, M.; Araki, K.; Taga, A.; Horiba, A.; Kawata, H.; Hirai, Y. Computational study on polymer filling process in nanoimprint lithography. *Microelectron. Eng.* **2011**, *88*, 2188–2191. [CrossRef]
23. Teyssèdre, H.; Landis, S.; Thanner, C.; Laure, M.; Khan, J.; Bos, S.; Eibelhuber, M.; Chouiki, M.; May, M.; Brianceau, P.; et al. A full-process chain assessment for nanoimprint technology on 200-mm industrial platform. *Nano Online* **2018**, *6*, 277–292. [CrossRef]
24. Teyssedre, H.; Landis, S.; Brianceau, P.; Mayr, M.; Thanner, C.; Laure, M.; Zorbach, W.; Eibelhuber, M.; Pain, L.; Chouiki, M.; et al. Rules-Based Correction Strategies Setup on Sub-Micrometer Line and Space Patterns for 200 mm Wafer Scale SmartNILTM Process within an Integration Process Flow. In Proceedings of the SPIE 10144, Emerging Patterning Technologies, 101440V, San Jose, CA, USA, 21 March 2017.
25. Ban, Y.H.; Bonnecaze, R.T. Minimizing filling time for ultraviolet nanoimprint lithography with templates with multiple structures. *J. Vac. Sci. Technol. B* **2021**, *39*, 012601. [CrossRef]

 *nanomaterials*

Article

Guiding Chart for Initial Layer Choice with Nanoimprint Lithography

Andre Mayer [1],* and Hella-Christin Scheer [2]

[1] Chair for Large Area Optoelectronics, School of Electrical, Information and Media Engineering, University of Wuppertal, Rainer-Gruenter-Str. 21, 42119 Wuppertal, Germany

[2] School of Electrical, Information and Media Engineering, University of Wuppertal, Rainer-Gruenter-Str. 21, 42119 Wuppertal, Germany; scheer@uni-wuppertal.de

* Correspondence: amayer@uni-wuppertal.de; Tel.: +49-202-439-1802

Abstract: When nanoimprint serves as a lithography process, it is most attractive for the ability to overcome the typical residual layer remaining without the need for etching. Then, 'partial cavity filling' is an efficient strategy to provide a negligible residual layer. However, this strategy requires an adequate choice of the initial layer thickness to work without defects. To promote the application of this strategy we provide a 'guiding chart' for initial layer choice. Due to volume conservation of the imprint polymer this guiding chart has to consider the geometric parameters of the stamp, where the polymer fills the cavities only up to a certain height, building a meniscus at its top. Furthermore, defects that may develop during the imprint due to some instability of the polymer within the cavity have to be avoided; with nanoimprint, the main instabilities are caused by van der Waals forces, temperature gradients, and electrostatic fields. Moreover, practical aspects such as a minimum polymer height required for a subsequent etching of the substrate come into play. With periodic stamp structures the guiding chart provided will indicate a window for defect-free processing considering all these limitations. As some of the relevant factors are system-specific, the user has to construct his own guiding chart in praxis, tailor-made to his particular imprint situation. To facilitate this task, all theoretical results required are presented in a graphical form, so that the quantities required can simply be read from these graphs. By means of examples, the implications of the guiding chart with respect to the choice of the initial layer are discussed with typical imprint scenarios, nanoimprint at room temperature, at elevated temperature, and under electrostatic forces. With periodic structures, the guiding chart represents a powerful and straightforward tool to avoid defects in praxis, without in-depth knowledge of the underlying physics.

Keywords: nanoimprint lithography; negligible residual layer; partial cavity filling; guiding chart; defect avoidance; hydrodynamic instabilities; T-NIL; UV-NIL; el-UV-NIL; el-T-NIL

Citation: Mayer, A.; Scheer, H.-C. Guiding Chart for Initial Layer Choice with Nanoimprint Lithography. *Nanomaterials* **2021**, *11*, 710. https://doi.org/10.3390/nano11030710

Academic Editor: Jose Maria De Teresa

Received: 10 February 2021
Accepted: 5 March 2021
Published: 11 March 2021

Publisher's Note: MDPI stays neutral with regard to jurisdictional claims in published maps and institutional affiliations.

1. Introduction

The basic idea behind nanoimprint was to propose a low-cost alternative for sub-micrometer lithography [1–3] without the need for highly sophisticated vacuum equipment. This intended application is expressed in its abbreviation, NIL (nanoimprint lithography). Typically, with lithography applications a thin polymeric layer on a hard substrate is imprinted. Meanwhile NIL has entered a much broader field of applications and is most often used for patterning the surface of polymeric substrates and foils. Surface structures may provide a specific wetting/de-wetting behavior for liquid phases [4]; surface structures also have a wide range of applications in optics, e.g., as gratings and anti-reflective, wave-guiding, or feedback structures [5–7]. In these cases, nanoimprint often does not involve thin layers on a hard substrate, and is therefore similar to hot embossing [8], a technique matured in the field of MEMS (micro-electro mechanical systems). Nanoimprint was reviewed with respect to a number of different aspects [9–19].

Here, we address nanoimprint as a lithography technique, employed to provide a local mask with isolated structures that define windows for further processing to the substrate underneath. With optical lithography, these windows result in a straightforward way by complete removal of the photoresist in exposed regions during development. However, with NIL, this is not naturally the case due to the residual layer, typically remaining with T-NIL (thermal NIL [1,10,20–23]) and with UV-NIL (ultraviolet-assisted NIL [2,3,24–27]) as well. Without specific precautions, a certain amount of imprint material always remains below the elevated stamp structures (see Figure 1a). With NIL it is common practice to remove this residual layer, e.g., by dry etching to obtain structures isolated from each other (Figure 1(a4)). An effective removal of the residual layer by dry etching asks for a high uniformity of the imprint to avoid a lateral loss of the masking structures during this step.

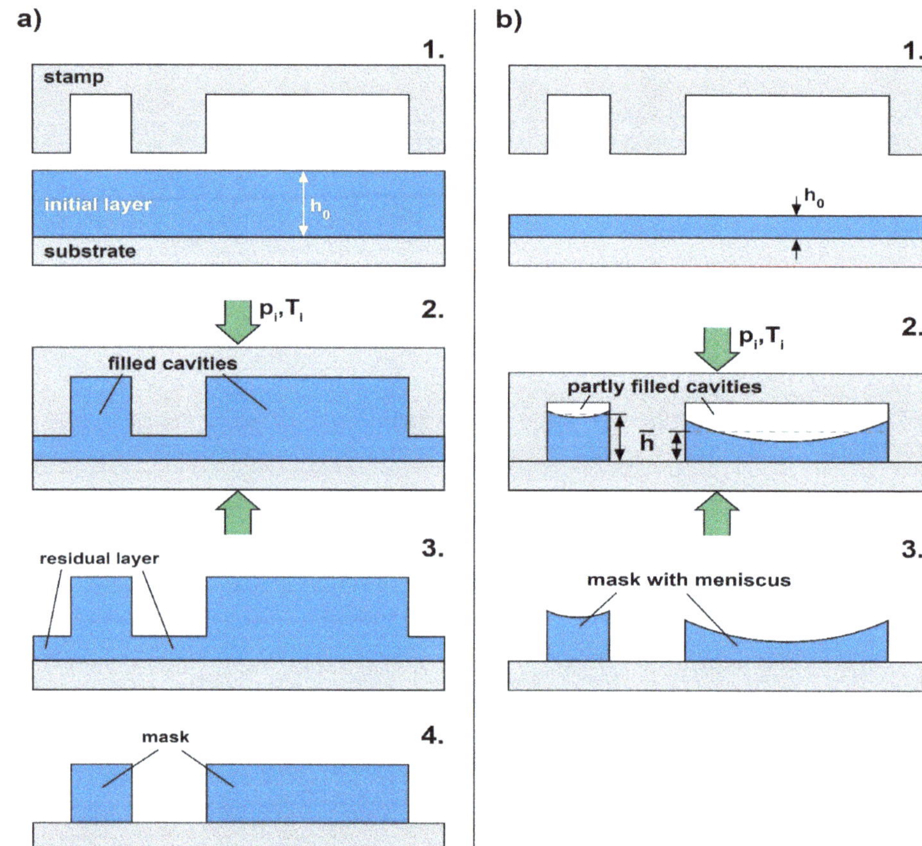

Figure 1. Imprint strategies with nanoimprint lithography (NIL) as a lithography process to provide a mask for the subsequent patterning of a substrate (p_i = imprint pressure, T_i = imprint temperature). (**a**) Conventional imprint with a residual layer remaining and filled cavities. (**b**) Concept of 'partial cavity filling' (idealized); all cavities are under-filled, and the polymer adopts a horizontal meniscus between the stamp sidewalls. 1. Initial situation with polymer height h_0. 2. Imprinted situation with/without residual layer and fully/partly filled cavities. 3. Polymeric structures after stamp removal, connected/isolated. 4. Polymeric structures after residual layer removal, with conventional imprint (**a**).

Some attempts were reported to imprint directly without a residual layer remaining to avoid such an intermediate etching step. With radiation-sensitive materials (curable materials, negative tone photoresists) just this property offers a way of lending itself to avoid etching. With UV-NIL, when the elevated structures of the stamp are provided with e.g., a metal, the residual layer is not cured and can simply be removed in a suitable solvent [28–30]. With T-NIL of photoresists, during flood exposure following imprint standing wave effects may simply be used to avoid curing of a thin residual layer [31]. In both cases the main limitation is a geometric one; the residual layer has to be thin and the stamp structures should be wider than the diffraction limit of the exposing radiation.

However, to achieve a thin residual layer is not trivial, in particular not over large areas. Due to the mechanical nature of nanoimprint, with UV-NIL and T-NIL as well, a uniform residual layer can only be obtained with periodic structures all over the stamp. With a non-periodic stamp either the structure height or the residual layer is non-uniform by itself [32]. The decisive factor in this context is the thickness of the stamp used. When the stamp is thick it is rigid enough to prohibit bending; then the residual layer is uniform but the structure height varies as some of the stamp cavities remain only partly filled. In contrast, with stamps that are thin or rather flexible, stamp bending may enable a complete filling of the cavities, however, then the residual layer becomes non-uniform [27,33–36]. With both non-uniformities, a varying structure height or a varying layer thickness, residual layer removal without structure loss is demanding [37].

Basically, any non-uniformity with locally varying lateral stamp geometries (the sizes of the cavities and the adjacent elevated stamp structures) results from a local filling of cavities [38]; any locally filled cavity impedes a further imprint in its neighborhood as the polymer retains its volume (compression negligible). In direct consequence, either the filling height (and thus the structure height) or the residual layer thickness (or both) become non-uniform. Hence, it is straightforward to avoid local cavity filling. One approach makes use of a stamp that locally provides additional cavity volume to equalize the residual layer, the so-called 'capacity equalized mold' [39]. In this approach, the stamp is prepared with differing cavity depths. Then, similar to the case of a rigid stamp, the residual layer is uniform, but the structure height varies. A simpler (and most efficient) approach with a constant cavity depth pursued in some groups is 'partial cavity filling', reported with soft stamps [40,41] and rigid stamps as well [29,42]. As sketched in Figure 1b, with 'partial cavity filling', the initial layer of the polymer is chosen so thin that the cavities remain under-filled with all stamp geometries involved. Again, the filling heights of the cavities and thus the structure heights then differ, but any residual layer can be largely avoided as (except for a few nanometers remaining [43]) the layer is imprinted through—basically the elevated stamp structures are in contact with the substrate at the end of the imprint (Figure 1(b2)). Stamp bending is avoided as well, as no cavity filling occurs. The method works well with stamps featuring a duty cycle (elevated stamp regions per area compared to cavities per area) that does not vary too much across the whole stamp, as it is the case with largely periodic structures.

Unfortunately, the partial filling of cavities implies an additional risk, namely the formation of defects via thermodynamic instabilities [40,44]. They are observed with viscosities low enough to allow their development within the actual processing time. With instabilities, the polymer within the cavities develops sub-structures with periodicities depending on the local filling height of the cavities [45,46]. The resulting defects are well-known to the experimentalist with T-NIL and UV-NIL as well [20,21,47–49]. It was proposed to make use of such defects for patterning of the polymer [48,50–54], however, as the control of the lateral dimensions and the periodicity over large areas is limited, such attempts are restricted to the lab scale. In order to apply nanoimprint as a lithography technique, defects caused by instabilities ultimately have to be avoided, they represent mask defects.

The present investigation aiming at residual layer-free imprint applies the strategy of 'partial cavity filling', where the polymeric structures feature a horizontal meniscus as sketched in Figure 1(b2,3). The key issue in this context is an adequate choice of the initial layer thickness to provide isolated structures. However, in order to achieve a continuous meniscus this choice has to consider the stamp geometries (stamp height/cavity depth, duty cycle) and their interplay with the contact angle that the polymer develops with respect to the stamp; this first issue is a mainly geometric one. Furthermore, to avoid defects developing with time due to instabilities, the viscosity of the imprint material at processing conditions (with UV-NIL at room temperature, with T-NIL at elevated temperature) also has to be included in the considerations; this second issue is a thermodynamic one. It is our intention to mitigate these complex geometric/physical relationships and to render them manageable for practical use. Therefore, we provide a tool for such an adequate choice of the initial layer thickness, a 'guiding chart'. This guiding chart indicates a window for defect-free processing by means of specific boundaries derived from the underlying physics. In our case, the boundaries are given in terms of the mean filling height and the cavity width, for a fixed height and duty cycle of the stamp. In contrast to attempts referring to soft molds [55,56], our approach is independent from the stamp material used and, in particular, considers thermally and electrically induced instabilities resulting from interaction of the imprint polymer with the ceiling of the stamp cavities.

To pave the way for a defect-free imprint with negligible residual layer thickness in praxis, it is essential to alleviate the complex physics of instabilities. To this end we will proceed as follows. (i) Where possible, we will work with generalized material properties and (ii) we will represent all theoretical relationships by diagrams from which the user is able to construct his own individual guiding chart (matching his specific imprint situation) in a straightforward way, without the need for a deep insight into physical relationships. We explain the procedure for constructing the guiding chart and discuss the processing window applying with T-NIL and UV-NIL, with and without electrostatic forces; by means of examples for these imprint techniques we propose an adequate choice of the initial layer thickness to avoid defects. Our simplifying approach is inspired by the characteristics of typical instability-induced defects observed with T-NIL. Starting with a guiding chart for periodic stamps with linear cavities, we will then broaden its applicability to cover periodic stamps with dot-like structures.

Our guiding chart provides an easily manageable tool to make use of otherwise rather complex physical relationships in praxis.

2. Materials and Methods

Typically, under processing conditions the materials to be processed via NIL are referred to as viscous liquids. Thus, their primary property is the viscosity (η), determining the time-scale of material response. Moreover, the surface tension/surface energy (γ) of the imprint material and the stamp as well are of interest as they define the contact angle that develops between the viscous liquid and its boundaries inside the partly filled cavities under equilibrium conditions, after a sufficiently long interaction time (e.g., 90 min with our material at 190 °C, to be on the safe side). The surface energy of the substrate is of minor importance here.

Experimental evidence is provided by T-NIL with PS (polystyrene) as the imprint material. Beyond the data for our specific polymer, we indicate general trends and correlations to enable assessing other imprint materials. Moreover, typical data with materials suitable for UV-NIL are included as the application of our guiding chart is independent from the specific NIL technique used.

As stated, the T-NIL experiments were performed with PS as the imprint material. The PS used here (350 kg/mol, Sigma Aldrich, Darmstadt, Germany) is well characterized with respect to imprint-related properties. Its glass transition temperature is 95 °C [57]. According to GPC-analysis (gel permeation chromatography), its molar mass values amount to M_n = 130 kg/mol (number average), M_w = 280 kg/mol (weight average), M_w/M_n = 2.1 (poly-dispersity) [58]; polymers with such a wide molar mass distribution are relatively easy to imprint, however their zero shear to shear-thinning transition is less pronounced than with polymers featuring a low poly-dispersity. Dynamic mechanical analysis was carried out in a plate-plate geometry to determine viscosity master curves; Table 1 summarizes the viscosities of our PS at specific processing temperatures. Please note that the correlation between temperature and viscosity depends on the molar mass of the polymer (η increases with increasing molar mass) [58,59]; in addition, it depends on the chemical structure of the monomer; PMMA (polymethylmethacrylate) of a similar molar mass has a higher viscosity than PS [60]. With a polymeric material given, temperature acts as the control parameter to tune the viscosity during processing, the material parameter of interest in praxis. Viscosities near 10^4 Pas and below are prone for instability-induced defects within 5 min of imprint time [57].

Table 1. Characteristic data of the imprint polymer used with our T-NIL experiments, PS, at typical processing temperatures (approximate values). η_p = zero shear viscosity, γ_p = surface tension, θ = contact angle to the stamp. Values for the viscosity and surface tension are taken from Refs. [58,60] and [61], respectively. The contact angle is determined from Young's equation, in good agreement with experiments [45]. The surface energy of the stamp used to determine θ amounts to $10/15/20$ mJ/m^2, referring to excellent/good/limited anti-sticking properties, respectively.

	25 °C	170 °C	190 °C	200 °C	250 °C
η_p/Pas	–	10^5	2×10^4	10^4	10^3
γ_p/mJ/m^2	41	33	31.5	30.5	27.5
θ/°	–	84/70/56	83/67/54	82/66/52	78/62/45

As further stated, the contact angle θ of the polymer within a partly filled cavity results from the surface energy/tension. The surface tension γ_p of the polymer decreases with increasing temperature [61]. Typical imprint materials have surface tensions of 30–40 mJ/m^2 at room temperature, decreasing by up to 10 mJ/m^2 at typical imprint temperatures with T-NIL. Surface tensions characteristic with our PS are also given in Table 1, together with the contact angle θ the polymer develops to the stamp walls. Crucial in this respect is the quality of the anti-sticking layer of the stamp. We assumed a surface energy of 10 mJ/m^2 and 15 mJ/m^2 (being largely independent from temperature) as typical with an anti-sticking layer of excellent or good quality that is required for a defect-free separation of the stamp from the imprinted sample. The third value considered, ≈ 20 mJ/m^2, refers to an anti-sticking layer of limited quality; this value indeed applies to a flexible stamp with a patterned surface layer from the elastomer PDMS (polydimethylsiloxane) as sometimes used with UV-NIL [17,19,40,62].

Table 1 indicates to what extent our characteristic material parameters vary with temperature. Typical viscosities effective with T-NIL are in the range 10^3 Pas $\leq \eta_p \leq$ 10^5 Pas, with all polymers; imprint materials differing from our specific PS (e.g., PMMA) may show these viscosities at a different temperature. Typical contact angles with T-NIL are in the range of $70° \leq \theta \leq 80°$.

For comparison, with UV-NIL viscosities are orders of magnitude lower during imprint at low pressure, in the range of 10^{-1}–10^2 Pas (e.g., epoxy silicone monomers [24] or other typical UV-NIL materials [26]); due to room temperature processing the surface tension and the contact angles are somewhat higher than with T-NIL ($\gamma_p \approx 30$–40 mJ/m^2, $\theta \approx 81$–$91°/66$–$77°/51$–$66°$).

For the T-NIL experiments, layers of different thickness were spin-coated to Si substrates from a toluene solution and dried (120 °C, 10 min, hotplate). Imprint was performed with Si stamps (2 cm $\times$ 2 cm) provided with an anti-sticking layer [63]. The stamps are almost completely patterned; they contain test structures, fields of lines, and spaces with different geometries. The parallel-plate imprint system used [64,65] is equipped with electrical heating; heating to the imprint temperature takes about 15–25 min; after about 5 min the glass transition temperature is reached. Cool-down to below glass temperature occurs within ≤ 2 min due to efficient water cooling. The imprint pressure is 100 bar, and a typical imprint time is 5 min. Inspection of the samples after sputtering with Au was done by scanning electron microscopy (SEM, S-FEG XL 30 S, Thermo Fisher Scientific, Waltham, MA, USA).

3. Typical Experimental Results

Figure 2 gives examples of typical imprint results obtained with T-NIL when the initial layer thickness h_0 is varied. Ideally, the imprinted polymeric structures should mirror the lateral stamp geometries, with a continuous horizontal meniscus at their top (see Figure 1(b2,3)). In praxis, when the initial layer thickness is not adequate, defects are observed, with polymeric structures smaller than the lateral dimensions of the cavities. In addition to defect-free structures, Figure 2 gives examples of defects characteristic of different cavity sizes. When the polymer is accumulated near an edge or corner the defect is geometry-related; when the polymeric sub-structures feature some periodicity, they result from instabilities. With narrow cavities, instabilities develop mainly along the length of the cavities; with wider geometries they may develop across the cavities as well. The examples mainly serve to illustrate nanoimprint under 'partial cavity filling' conditions to provide residual layer-free structures for lithography.

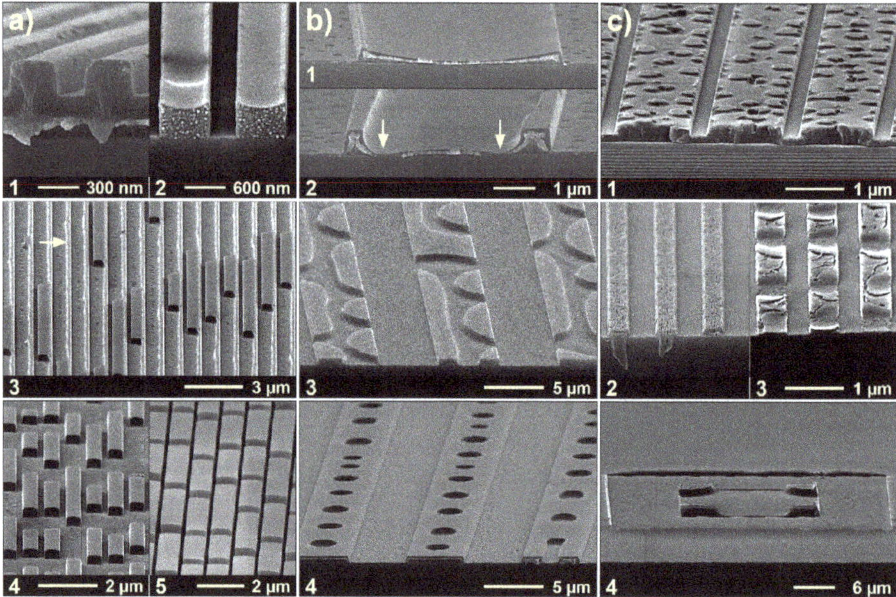

Figure 2. Examples obtained with T-NIL at differing initial layer thickness h_0. If not stated otherwise in the text, the imprinted material is PS, and the imprint temperature is 190 °C. (**a**) Examples with narrow cavities (w = 300–800 nm). (**b**) Examples with wider cavities ($w \approx 5$ μm). (**c**) Miscellaneous examples see text. The initial layer thickness increases from (**a3**) to (**a5**) and from (**b2**) to (**b4**).

Most of the examples were obtained at an imprint temperature of 190 °C, where the viscosity of the PS used is low enough to allow the polymer that is squeezed into the cavity from the sides to adopt an equilibrium state within the partly filled cavities. This equilibrium state is characterized by a typical contact angle of the polymer towards the stamp (70°–80°). Though the examples were obtained with T-NIL only, they are similarly typical of results to be obtained with UV-NIL at room temperature, merely at much lower viscosities (and low driving forces/pressures) [2,3,24–26,40,66].

Figure 2(a1) shows an example with complete filling of the cavities, all other examples refer to 'partial cavity filling' conditions. Column (a) gives examples with narrow cavities, $w \approx 300$–800 nm; column (b) gives examples with wider cavities, $w \approx 5$ μm, and column (c) refers to differing specific situations. Most of the examples verify the existence of instabilities of the polymer within the cavity as some periodicity can be identified.

When the stamp cavities are completely filled, a residual layer typically remains below the imprinted structures, as clearly visible with (a1). When the residual layer is high compared to the structure height and when it is non-uniform (as it would be the case with locally differing duty cycle), residual layer removal by dry etching is challenging. In contrast, structures without a residual layer could serve as a mask directly, without an intermediate etching step.

With 'partial cavity filling' the most regular shape is the one of a continuous horizontal meniscus between the stamp sidewalls as documented in Figure 2(a2,b1), taking the example of a narrow and a wide cavity. When the masking height (the center height of the meniscus in the cavity) is sufficient, the structures with a horizontal meniscus are those that are well-suited to serve as a mask for etching of the substrate (e.g., a2); this is why structures with a continuous horizontal meniscus (see Figure 1(b2,3)) are the target structures for lithography applications of NIL (the masking height required depends on the mask selectivity of the etching process applied). With a similar mean filling level a sufficient masking height in the cavity center is easier to obtain with a smaller cavity, due to the spherical shape of the meniscus (see e.g., Figure 2(a2) compared to Figure 2(b1)).

Figure 2(b2–4) gives examples, where the meniscus formed involves the stamp ceiling rather than the stamp sidewalls. This occurs when, during meniscus formation, the contact line of the polymer at the sidewalls reaches the stamp corner. Then a rim forms all along the sidewalls, where the polymer fills the stamp cavity locally to its full height (b2); the rim increases in width when the initial layer thickness increases. As the formation of the rim results in a withdrawal of polymer from the neighborhood (see arrows) the masking height there may become small locally and result in de-wetting, precluding the use of such structures for lithography in praxis. Though (b2) results from purely geometric reasons, the fluctuations of the rim along the length of the cavity already indicate the onset of instabilities. Figure 2(b3,4)) illustrates the situation when the initial layer height is increased (compared to b2), thus increasing the mean filling level in the cavity. The periodicity of alternating filled and empty regions observed now is typical of fully developed instabilities, applying to the width and the length of the cavity. Again, these structures are unsuitable for lithography as they do not mirror the lateral stamp geometries. At relatively high filling level (b4) often a series of holes remains along the center of the cavity. If these holes were surface-near only so that a sufficient height of polymer remained below them for subsequent etching, such structures could serve as a mask. However, as the contact angle is below $90°$ with nanoimprint, these holes widen to the bottom; in addition, some randomness exists. Therefore, also polymeric structures similar to (b4) are too risky for lithography applications.

With narrow cavities (a3–5), the transition from a purely geometric effect, the polymer-stamp contact at the edge of the stamp cavity, to fully developed instabilities are less obvious. The cavities are either completely filled or empty, alternating along the length of the cavity, only. With a high mean filling level, the 'on-off'-ratio is high (a4,5); (a3) is a rare example where the rim along the sidewalls of the stamp is still visible (see arrow), between regions of complete stamp filling (the stamp was not parallel to the substrate). Of course, these structures are far from being qualified for lithography purposes, where masking of the substrate along the complete length of the lines would be asked.

Figure 2(c1) refers to a specific situation where the imprint still shows a residual layer but the cavities are not completely filled (the stamp was held at a certain distance due to pattern size effects [67]. Within the small gap between the polymer surface and the stamp ceiling, periodic instabilities are visible. These small differences in polymer height are not restrictive for lithography, even when some residual layer has to be removed first, as it is the case here.

Figure 2(c2,3) refers to a thermal imprint under electrostatic forces. Here a lamellar block-copolymer was imprinted (PS-PMMA) under 'partial cavity filling' conditions, where a voltage was applied to induce a vertical phase separation in the block-copolymer. The situation in (c2) refers to a low imprint temperature (170 $°$C); instabilities did not occur.

In contrast, instabilities along the cavities are fully developed when imprinting at high temperature, 210 °C (c3). In both cases the PMMA-part of the block-copolymer was removed, so that the PS-component remained, only.

Finally, Figure 2(c4) refers to a wide cavity of square shape (with a thick wall), located in the center of the micrograph. The polymer has filled the stamp cavity at the inner edges of the cavity only; any periodicity from instabilities is not visible. This results from a mainly geometric effect. Similar to the situation with a linear cavity (b2), a 2-dimensional cavity features the highest polymer level at the inner edges (due to meniscus formation in two directions). There, the polymer filled the cavity in full height, under withdrawal of material from the rest of the cavity. Again, suitability for lithography is precluded as the rest of the cavity is not masked.

From these examples observed in praxis we draw the following conclusions. With 'partial cavity filling', the meniscus at the polymer surface plays a central role for formation of intact or defective structures. The meniscus is controlled by the contact angle θ the polymer develops to the stamp. As long as a continuous horizontal meniscus forms between the stamp sidewalls, the polymeric structures obtained are well-suited for lithography purposes when their minimum height is sufficient for subsequent etching. This is the case when the polymer develops its contact angle to the sidewalls of the stamp, at mean filling levels of adequate height.

However, at a filling level too high the meniscus touches the ceiling of the stamp at the edges of the cavity. This results in a sort of 'mode jump', from a horizontal meniscus to a vertical meniscus. Now the polymer tends to develop its contact angle to the horizontal ceiling of the stamp cavity (and to the substrate). As a consequence, a reconfiguration of the polymeric liquid within the cavity occurs. This may lead to local defects where the substrate becomes exposed (de-wetting, see arrows in b2), rendering these polymer structures inapplicable to lithography, similar to the structures in b4.

As will be addressed in part 4, instabilities may induce this mode jump at even lower filling levels. With instabilities, the polymer layer breaks down into more or less periodic sub-structures, characterized by a local filling of the stamp cavity to its full height alternating with a vanishing height. Depending on the lateral dimensions of the cavity and the characteristic periodicity, the reconfiguration of the polymer within the cavity may affect both geometries, its width, and its length. Though fascinating from a physical perspective instability-induced structures are less useful in praxis as some randomness over large areas cannot be avoided. In particular they are fully inappropriate for lithography purposes with nanoimprint as they do not mirror the lateral stamp geometries.

Generally, the onset of instabilities at inadequate filling levels limits the use of 'partial cavity filling' for nanoimprint with negligible residual layer. This holds with a mean filling level too high (as exemplified in Figure 2) and too low as well; the latter may lead to de-wetting in the center of wide cavities.

As a consequence, only structures with a continuous horizontal meniscus are able to mask the substrate. The process window to be defined thus refers to the regime of the initial polymer height h_0 where such a horizontal meniscus forms when a stamp of height H and with certain geometries of the elevated structures (s) and the cavities (w) is imprinted into the thin layer. The intention of nanoimprint with 'partial cavity filling' is to provide such structures without a residual layer, isolated from each other. Its realization requires an adequate choice of the initial layer thickness.

In the following, we will address the limitations imposed on 'partial cavity filling'. We will start with purely geometric limitations (similar to b2 and c4) and present a preliminary guiding chart to identify a basic processing window for defect-free imprint (part 4), indicating a maximum cavity width allowed. Beyond imprint-specific aspects, application-specific aspects will also be addressed. This geometric guiding chart will then be modified to include instabilities (part 5). Finally, the construction of a specific guiding chart for practical application is addressed (part 6). Consequences for the choice of the initial layer are drawn and potential measures to widen the processing window are indicated.

4. Geometric Processing Window

A basic, purely geometric processing window can be identified from a guiding chart in a plane that is spanned by the mean filling height in the cavity, $\bar{h}$, and the cavity width, w. Further geometry parameters and material parameters involved are the stamp height H, the size of the elevated stamp structures s (or rather their duty cycle s/w), the initial layer thickness h_0 and the contact angle θ between the polymer and the stamp. According to the concept of 'partial cavity filling' we assume that all the polymer available has been squeezed into the cavities and that a horizontal meniscus has formed between the stamp sidewalls, the target shape for lithography applications. With T-NIL, volume conservation holds in good approximation for a polymer above its glass transition, similar to the situation with (almost) liquid low viscosity resins used in UV-NIL. Furthermore, as all geometries are small, the effect of gravity can safely be neglected, and the horizontal meniscus formed within a cavity has a spherical or cylindrical shape.

This purely geometric guiding chart is already adequate to discuss the lower limit of the window for defect-free processing (at low mean filling levels $\bar{h}$), under practical, application-specific aspects. A correct identification of the upper limit (at high mean filling levels $\bar{h}$), will require a modification, namely the consideration of instabilities, as addressed in paragraph 5.

For simplicity we develop the guiding chart in the first instance by taking a frequently met example, the imprint of a linear, 1-dimensional grating with a certain duty cycle s/w. We assume the stamp to be patterned over its whole area, so that edge effects are avoided [34,38]. As an extension, the basic relationships for a guiding chart referring to a 2-dimensional grating are given in Appendix B.

4.1. Construction of the Geometric Guiding Chart

Figure 3 gives details of the quantities relevant to describe the structures obtained with a linear grating under 'partial cavity filling' conditions. Figure 3a illustrates the cross-sectional situation. The polymer available per period within the initial layer (Figure 1) is $h_0(w + s)$, which leads to a mean filling height $\bar{h}$ (see Figures 1 and 3) in the periodic cavities of

$$\bar{h} = h_0\left(\frac{w+s}{w}\right) = h_0\left(1 + \frac{s}{w}\right) \tag{1}$$

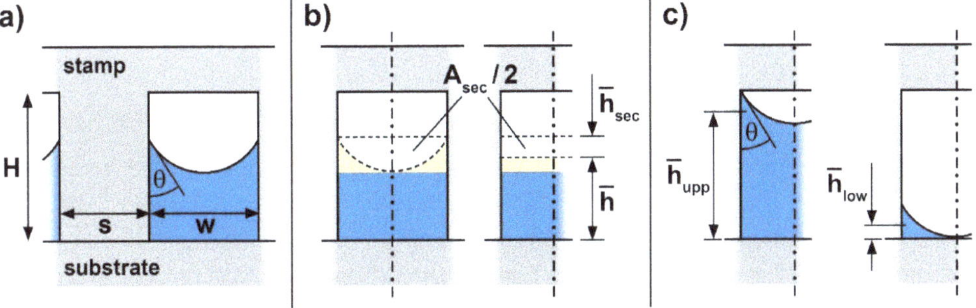

Figure 3. Definitions in view of the specification of a purely geometric processing window; cross-section through periodic linear structures. (**a**) Imprint situation: Stamp with geometries s, w, and H; within the partly filled cavities the polymer surface adopts a meniscus according to its contact angle θ; (**b**) partitioning of cross-sectional areas (see text); A_{sec} refers to the circular section as indicated (mean filling height $\bar{h}$, mean height of circular section $\bar{h}_{sec}$). (**c**) Upper and lower boundaries: Polymer meniscus touches the ceiling of the stamp cavity at the corners (mean polymer height $\bar{h}_{upp}$) and the substrate in the center of the cavity (mean polymer height $\bar{h}_{low}$), respectively.

We identify this mean filling height as the parameter of interest as it is directly related to the initial layer thickness h_0 to be chosen experimentally and the duty cycle (here s/w).

Therefore, our guiding chart, Figure 4, maps the mean filling height, $\overline{h}$, for a range of cavity widths, w.

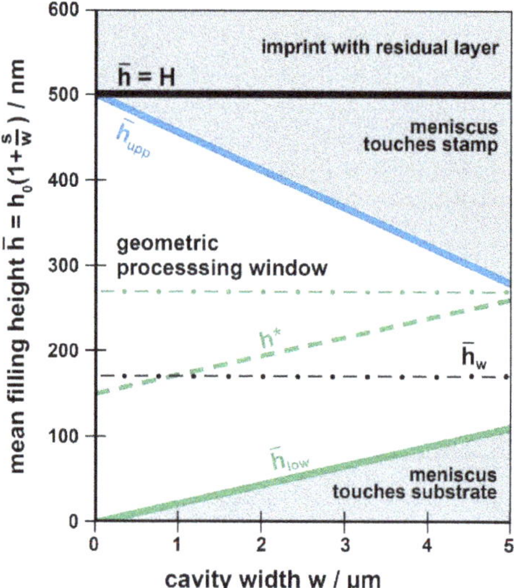

Figure 4. Example of a guiding chart for defect-free imprint with negligible residual layer, considering purely geometric limits (mean filling height $\overline{h}$ as a function of cavity width w for a contact angle of $\theta = 75°$ and a stamp height of $H = 500$ nm). The geometric processing window considering imprint-related issues (bright) is limited at its top and its bottom. The following boundaries and heights are indicated: $\overline{h} = H$ Exact filling of the stamp cavities without residual layer. $\overline{h}_{upp}$: Meniscus in stamp cavity reaches stamp ceiling at edges. $\overline{h}_{low}$: Centre of the meniscus touches substrate. $\overline{h}_w$: Optimum choice of mean filling height with mixed cavities (dash-dotted). h^*: Additional boundary when a minimum polymer height is required (dashed, see text).

When this mean filling height $\overline{h}$ equals the stamp height H the cavities are completely filled without any residual layer. This is our first boundary for the guiding chart,

$$\overline{h} = H; \tag{2}$$

it separates the regime with a residual layer remaining ($\overline{h} > H$) from the one of 'partial cavity filling' ($\overline{h} < H$), without residual layer.

Within the regime of 'partial cavity filling' two further geometric boundaries exist, as illustrated in Figure 3c, an upper one, where the meniscus just meets the edge of the stamp (mean filling height $\overline{h}_{upp}$), and a lower one, where the center of the meniscus touches the substrate (mean filling height $\overline{h}_{low}$).

Thanks to the cylindrical shape of the meniscus in our linear cavities the respective mean filling heights can easily be calculated based on the area of a circular section A_{sec} as indicated in Figure 3b [45]. The mean height of such a circular section, $\overline{h}_{sec}$, when expressed by the parameters θ and w (the parameters involved here) amounts to

$$\overline{h}_{sec} = \frac{A_{sec}}{w} = \frac{\pi\,(1-\theta/90°) - \sin 2\theta}{8\,\cos^2\theta} \cdot w. \tag{3}$$

This relationship helps to determine the upper and lower boundaries, as depicted in Figure 3c. The upper limit (where the polymer meniscus touches the stamp ceiling at the corners of the cavity) is reached at a mean filling height of

$$\overline{h}_{upp} = H - \overline{h}_{sec} = H - m_{upp} \cdot w, \tag{4}$$

whereas the lower limit (where the polymer meniscus touches the substrate in the center of the cavity) is given by

$$\overline{h}_{low} = \frac{1 - \sin\theta}{2\cos\theta} \cdot w - \overline{h}_{sec} = m_{low} \cdot w. \tag{5}$$

This lower limit is also reported by other groups [55].

These two geometric boundaries depend linearly on the cavity width w; the respective slopes m_{upp} and m_{low} are defined by the contact angle θ between the polymer and the stamp which, for a given imprint situation, is constant (characteristic values see Table 1).

To simplify the construction of the guiding chart for users the slopes of the two boundaries, m_{upp} and m_{low} (Equations (4) and (5)), are plotted in Figure 5 as a function of the contact angle θ. Two linear approximations to these curves are indicated, representing the slopes with high contact angles, $\theta \geq 50°$. These approximations are well-suited for contact angles that are typical of nanoimprint (see Table 1), the non-linear part of the relationships being meaningless for practical applications. The approximations are given by the numerical relationship

$$m_{upp} \approx 0.27 - 3 \cdot 10^{-3} \frac{\theta}{\deg} \approx 2\, m_{low} \tag{6}$$

and differ simply by a factor of 2. In most cases, Equation (6) is appropriate to determine the values for the two limiting slopes, as an alternative to reading the approximate value from Figure 5.

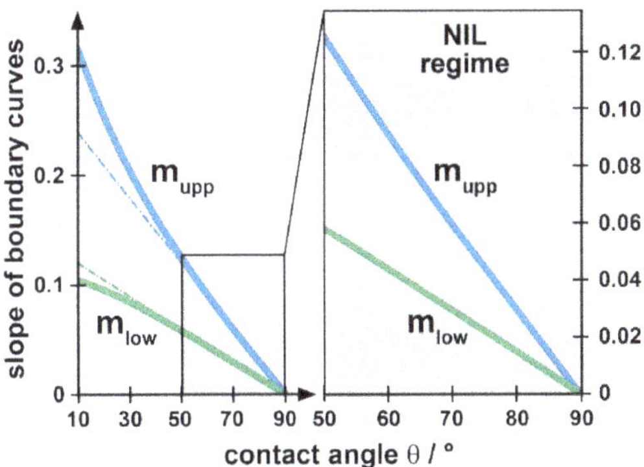

Figure 5. Non-dimensional slopes of the upper (blue) and lower (green) boundary for the formation of a horizontal meniscus in a linear cavity as a function of the contact angle θ of the imprinted polymer with respect to the stamp. With high contact angles the linear approximations are adequate (dash-dotted).

The geometric guiding chart (mean filling height $\overline{h}$ over cavity width w) is constructed by choosing a range of cavity widths of interest and by indicating the height of the stamp to be used, H, as an upmost limit for $\overline{h}$. Then the two straight lines representing the upper and lower boundaries, $\overline{h}_{upp}$ and $\overline{h}_{low}$ are drawn, taking the slopes m_{upp} and m_{low} from Figure 5 or Equation (6). The region between $\overline{h}_{upp}$ and $\overline{h}_{low}$ is the available processing window, under purely geometric, imprint-related limitations. Obviously, the cavity width sets an absolute limit for nanoimprint under 'partial cavity filling' conditions for lithography purposes, with a specific height H of the stamp.

4.2. Discussion of the Geometric Guiding Ghart

Figure 4 gives an example of a preliminary, purely geometric guiding chart. As an example, we take a contact angle of $\theta = 75°$ and consider cavities w of up to 5 µm width. We assume a stamp of 500 nm in height, as indicated by the horizontal line at $\bar{h} = H$. The slopes m_{upp} and m_{low} from Figure 5 ($m_{upp} \approx 0.045 = 45$ nm/µm $= 2\, m_{low}$) are used to draw the boundaries $\bar{h}_{upp}$ and $\bar{h}_{low}$. Note that precision of the values is not an issue for constructing the guiding chart in praxis; the intention behind it is just to become familiar with the limitations existing with 'partial cavity filling'. Any choice of initial layer thickness too near to any limit does not make sense in praxis, as a lot of the parameters required for constructing the guiding chart are known as approximate values only. In addition, small local variations of the geometries (e.g., stamp roughness) may lead to a random fluctuation of the boundaries.

4.2.1. Imprint-Specific Issues

Only within the region between $\bar{h}_{upp}$ (blue line) and $\bar{h}_{low}$ (green line) a vanishing residual layer together with a horizontal meniscus in the cavities can be obtained. With increasing cavity width, the window of allowable mean filling heights $\bar{h}$ narrows as the risk that the meniscus touches the substrate or the ceiling of the stamp cavity increases. For any cavity width, the range of allowed initial layer thicknesses h_0 (corresponding to $\bar{h}$) can easily be derived from the stamp duty cycle according to Equation (1).

Generally, the wider the cavities of a linear grating (at a duty cycle $s/w = $ const) the more accurate the initial layer h_0 has to be chosen to avoid defects. With varying cavities, the 'most save' initial layer is the one resulting in a mean filling height of $\bar{h}_w \approx H/3$, due to the relationship of Equation (6); it provides defect-free imprint with vanishing residual height over the widest range of cavity widths.

4.2.2. Application-Specific Issues

So far, imprint-specific issues were addressed. Of course, choosing the initial layer with the help of the guiding chart cannot be based on imprint-related issues only; however, application-specific issues have to be considered, too. Typically, with NIL as a lithography process, a mask shall be provided for subsequent etching. When an intermediate lift-off process is intended (e.g., to invert the tone) the main issue is that the polymeric structures within the cavities are without holes. Then, theoretically, the processing window already discussed is usable, with a lower limit $\bar{h}_{low}$. However, to be on the safe side and to ensure that the substrate is masked even with local non-uniformities it is advised to prescribe a certain minimal polymer height, h^*, in praxis. Moreover, when the imprinted structures shall provide the etching mask directly, this lower limit is defined by the etching process. As dry etch selectivity is limited, the minimum polymer height required to mask the substrate depends on the dry etching process. Its value is known by the user, only; any generalizing assumption is not possible. However, with h^* at hand this additional limit can be integrated into the guiding chart. As $\bar{h}_{low}$ indicates a vanishing polymer height in the center of the cavity, this line has to be shifted upwards by h^*. This additional practical boundary further narrows the suitable processing window from its bottom. As an example, a safety margin (or etch-induced limit) is indicated in Figure 4 assuming an arbitrary value of $h^* = 150$ nm (which would for example not be reached with the example in Figure 2(b1), despite the continuous meniscus). With this example cavities of $w \geq 5$ µm are critical.

Combining the processing window for defect-free imprint with this additional, application-related boundary clearly indicates, up to which cavity size partial cavity filling is suitable with periodic, linear structures to provide a polymeric mask for the specific dry etching process to be used. As a consequence, under such conditions the regime of higher filling levels, $\bar{h} > h^*$, is of primary practical interest within the defect-free window.

Of course, the filling height providing the best choice for a wide range of cavities then shifts upwards, accordingly (h^*).

4.2.3. Applicability

In the present form, the guiding chart, Figure 4, is applicable when just geometric factors represent the limits. This is the case when imprinting materials of very high viscosity, in the range of $\eta \approx 10^6$ Pas, as it would apply to T-NIL at a very low temperature. With lower viscosities, $\eta \leq 10^4$ Pas (a more typical processing regime with T-NIL), the polymer layer may suffer from instabilities developing within characteristic processing times, resulting in additional masking defects. UV-NIL may be even more prone to instabilities, due to the typically low viscosities. These instabilities are addressed in the next paragraph.

5. Processing Window with Instabilities

Before revising the geometric processing window with respect to instabilities a basic view of the common knowledge is provided here. As our focus is on a practical manage-ability, the complex relationships existing with instabilities are generalized and presented by means of diagrams enabling the user to exploit such knowledge for his own specific processing situation. To explain the origin of the limitations induced by instabilities the physical background is summarized in short; an understanding of this background is not required, neither for constructing nor for using the guiding chart to identify the defect-free processing window in praxis. Readers interested primarily in the use of the guiding chart may skip part 5 (5.1 refers to the general physics underlying, and part 5.2 refers to the specific parameters entering the guiding chart). For the construction of the guiding chart, just the results will be used, namely Figures 6 and 8, together with the definition of the respective process parameters of relevance, Equations (10), (12) and (14).

5.1. Physical Background

Basically, the surface of a polymeric liquid, when near an interface, is subjected to interaction with this interface. With a thin film it simply is the interaction with the substrate, which may lead to a de-wetting of the film [68]. In the case of nanoimprint, the situation is more complex, as there may be an interaction that exists with the substrate, in addition to the stamp [69]; this is the situation of the polymer within a cavity under 'partial cavity filling' conditions. In particular, when an instability due to interaction with the stamp occurs, a small undulation of the polymer height will grow. When the stamp ceiling is reached somewhere after some time, this local stamp contact results in a reconfiguration of the polymer within the cavity to minimize the free energy, the faster the lower the viscosity. Such a reconfiguration may lead to a de-wetting between the contact points. Of course, a de-wetting to the substrate is also possible without stamp contact; then the interaction with the substrate is dominating as characteristic of very thin layers. Typically, a 'linear stability analysis' [52,70–72] is performed to identify those undulations that grow fastest, the only ones to be observed in the experiment.

5.1.1. Driving Forces with NIL

Driving forces for instabilities in general are numerous [73]. With nanoimprint, the most important drivers are addressed by Schäffer [71,72,74,75]—our treatment is based on his work. These drivers are van der Waals forces, forces due to temperature gradients and electrostatic forces. Van der Waals forces exist in any imprint configuration; temperature gradients are specific of non-isothermal processes; electrostatic instabilities require the application of an external voltage.

Independent from the imprint technique used, T-NIL or UV-NIL, van der Waals forces are always active to drive instabilities, in particular with thin fluid films [76]. With T-NIL, there will be instabilities from temperature gradients as well. Temperature gradients are most obvious with imprint systems featuring single-sided heating [77–79]. However, systems serving for iso-thermal processing also often feature temperature gradients that are less obvious. With parallel-plate systems, temperature differences of 1–2 °C are typical during imprint. Though small, we found that such temperature differences may already induce instabilities [45]. Furthermore, during the heat-up and cooling phase temperature differences between the heating plates may be even larger; 10 °C are not unusual [46]. At temperatures above the glass transition of the imprint polymer this heating/cooling phase has to be included when considering instabilities with T-NIL. We anticipate that with T-NIL, barely any system exists where instabilities from temperature gradients can safely be ignored. The system used for our experiments features a temperature gradient of about 10 °C during heat-up and a temperature gradient of 1–2 °C during imprint [46]. The temperature gradients with T-NIL are system-specific.

Electrostatic forces may also be system-specific; this is the case when they serve to provide the pressure for imprinting [80]; typically, this works with curable materials of low viscosity at room temperature as generally used with UV-NIL; we will refer to this technique as 'el-UV-NIL'. In this case, the existence of electrostatic forces is obvious; however, due to their coupling with the imprint pressure a free choice of their size is impeded. Furthermore, electric fields may be applied intentionally during T-NIL to induce certain physical effects, e.g., the phase separation of a block copolymer [81,82]. Then both, electrostatic and thermal instabilities may occur, the largest one dominating the situation; we will refer to this as 'el-T-NIL'. Moreover, electric fields may also exist unintentionally, e.g., when T-NIL is performed via current-induced heating of the stamp itself [77,78]; whether or not electric fields are present in such a case depends on the implementation of the electrodes and their grounding situation. Such unintended co-action of temperature differences and electric field ranks again as 'el-T-NIL'.

To address all these aspects with NIL, our analysis will cover van der Waals forces, temperature gradients, and electrostatic forces.

5.1.2. Stability Analysis with NIL

To identify the basic relationships, we follow the conventional procedure and perform a linear stability analysis. Here it refers to a polymeric layer of mean height $\bar{h}$ within a stamp cavity of height H, according to Figure 1(b2) and Figure 3b. The following simplifications are made. (i) We separate the imprint from the stability analysis; thus, we assume the instabilities to develop when the polymer has already been squeezed into the cavities. (ii) We assume that an equilibrium is reached within the processing time. (iii) We ignore the lateral meniscus in the cavity and work with the mean filling level $\bar{h}$ instead. (iv) We consider instabilities in vertical direction only; due to some randomness in the experimental conditions (e.g., stamp roughness [83]), an exact consideration of lateral boundaries is less meaningful.

We found that this simplified analysis, in combination with the geometric analysis already discussed, is appropriate to understand all typical phenomena observed experimentally with instabilities during nanoimprint. In particular, we make use of the basically periodic nature of the instability phenomenon to apply it to the lateral cavities. With linear cavities of a few micron width, the experiments (see Figure 2(b2)) indicate that instabilities start from the sidewalls of the stamp and then develop in the third dimension, along the length of the respective cavity, during the reconfiguration phase.

Furthermore, as our aim is to draw practical conclusions from such a stability analysis for nanoimprint, we restrict the presentation here to those equations/correlations that are urgently required to explicate the actual proceeding. Details of the analysis are given in Ref. [45] and in parts in Ref. [46]. To assign the equations given in Ref. [46] to the actual situation the initial layer thickness h_0 and the gap height d (distance between polymer surface and stamp ceiling) have to be replaced by the respective actual quantities, the mean filling height in the cavities, $\bar{h}$, and the height of the stamp structures, H.

Mathematical treatment starts from a polymer of mean height $\bar{h}$ located between the substrate and the stamp (at a distance H). Its surface, the polymer/air-interface, is subjected to interactions with other nearby interfaces, the polymer/substrate and the air/stamp-interface. These interactions are characterized by a respective pressure, p, which may 'stabilize' or 'destabilize' the layer. Destabilizing pressures (inducing instabilities by amplifying fluctuations) result from the forces identified as drivers, van der Waals forces (p_{vdW}, between the polymer surface and the stamp or substrate), thermal forces (p_{th}, when a temperature gradient exists between stamp and substrate), and electrostatic forces (p_{el}, when a potential difference U is effective between stamp and substrate). These 'destabilizing' pressures ($p_{des} = p_{vdW} + p_{el} + p_{th}$) are counteracted by the surface tension of the viscous layer via the Laplace pressure p_{La} (that tries to hold the polymer-air interface as flat (and thus, as small) as possible to minimize the respective energy); thus, p_{La} acts as the major 'stabilizing' pressure ($p_{La} = p_{stab}$) that damps instabilities.

With imprint, the polymer volume remains conserved, with T-NIL above the glass transition of the polymer and with low viscosity liquid resins used for UV-NIL as well. Therefore, under the action of pressures the viscous polymer layer is described by the continuity equation, here a differential equation for the polymer height $h(x,t)$. With our 2D-problem (linear grating), solutions for this differential equation are harmonic in lateral direction x (with wave numbers k_i) and exponential in time t (with time constants τ_i). With one of these harmonic solutions (i) the fluctuating polymer height follows

$$h_i(x,t) \;=\; \overline{h}\left(1 + \delta \cdot \cos\left(\frac{2\pi x}{k_i}\right)\exp\left(\frac{t}{\tau_i}\right)\right), \tag{7}$$

with δ some small quantity ($\delta << 1$) defining the undulation of the initial amplitude of the polymer height over its mean value, $\overline{h}$.

As just the fastest growing undulation shows up in the experiments, the solution with the smallest time constant is the only one of practical interest. Based on Equations (3) and (4) of ref [46] this smallest time constant, τ, is defined by the mean layer thickness $\overline{h}$ and the change of the destabilizing pressure with layer thickness, according to

$$\frac{1}{\tau} = \frac{\overline{h}^{3}}{12\eta_p\gamma_p}\left[\left(\frac{\partial p_{des}}{\partial h}\right)^{2}\right]_{h=\overline{h}}, \tag{8}$$

with η_p and γ_p the viscosity and the surface tension of the polymer under processing conditions.

When, after a certain interaction time t_p (characteristic for the process), the maximum of the fastest growing undulation reaches the ceiling of the stamp cavity ($h_{\max}(t_p, \tau) = H$) a polymeric bridge forms. Combining Equations (7) and (8) under such conditions results in the relationship

$$0 = P_0 \cdot \frac{\overline{h}^{3}}{12}\cdot\left[\left(\frac{\partial p_{des}}{\partial h}\right)^{2}\right]_{h=\overline{h}} - \ln\left(\frac{H - \overline{h}}{\delta\cdot\overline{h}}\right), \tag{9}$$

with

$$P_0 = \frac{t_p}{\eta_p\gamma_p}. \tag{10}$$

P_0 is a basic processing parameter. The quantity dominating its size is the viscosity η_p, which may vary over six decades with the different imprint techniques. It will be used later on to define a process-specific parameter that is characteristic of thermal and electrostatic instabilities.

Equation (9) is an implicit relationship between the mean polymer height in the cavity, $\overline{h}$, and a given stamp height, H, with any destabilizing pressure of interest; the basic processing parameter P_0 being constant with a specific imprint situation. A solution $\overline{h}\,(H)$ implies that the right-hand side of Equation (9) equals zero; then the polymer level $\overline{h}$ is high enough that the undulations just touch the stamp within the interaction time, bridging the initial gap, $(H - \overline{h})$. This is the boundary looked for that separates the defect-free from the defective regime. A choice of $\bar{h}$ below the boundary does not result in instabilities, but beyond it does.

Calculating the boundary means to find a solution $\bar{h}\,(H)$ of Equation (9). Generally, the destabilizing pressure of interest is inserted in Equation (9) and the implicit mathematical relationship is solved for its null for pairs of variates of $(\bar{h}, H)$, taking the specific material and process parameters into account as explained below with thermal and electrostatic instabilities.

5.2. Modification of the Guiding Chart with Instabilities

In the following, the three destabilizing forces (van der Waals, thermal, electrostatic) are each addressed to identify the respective boundaries that further limit the processing window with instabilities, as exemplified in the modified guiding chart, Figure 6. Its construction is based on the limits already discussed with the purely geometric processing window, Figure 4.

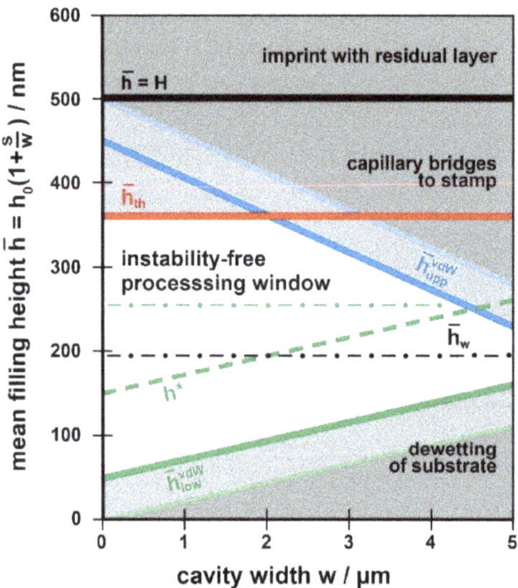

Figure 6. Example of a modified guiding chart for defect-free imprint with negligible residual layer similar to Figure 4, but additionally considering instabilities. The processing window (bright) is further reduced by boundaries indicating instabilities, namely $\bar{h}_{upp}^{vdW}$ and $\bar{h}\,_{low}^{vdW}$: Instabilities due to van der Waals forces. $\bar{h}_{th}$: Instabilities due to temperature differences (with instabilities due to electrostatic forces $\bar{h}_{th}$ would be replaced by $\bar{h}_{el}$).

To provide practical access to this complex physical regime we will generalize the results as far as possible. This is facilitated by the fact that within the typical parameter range the material properties (dielectric, thermal) are not crucial for the main findings. With thermal and electrostatic instabilities, we will identify a single process-specific parameter (P_{th}, P_{el}), combining the basic processing parameter P_0 (Equation (10)) with the external control parameter applying, the temperature difference ΔT or the voltage applied U. The solutions of the problem are presented as a graph. The limiting polymer height can be read from the graph with this single process-specific parameter at hand, for any stamp height.

5.2.1. Van der Waals Forces

Van der Waals forces between plane interfaces are proportional to the reciprocal of the respective distance to a power of three; accordingly they affect thin layers or narrow gaps only. The material characteristics enter the relationship via the respective Hamaker constant being dielectric in nature [76,84]. The situation under typical imprint conditions is addressed in detail in Refs. [45,46]. Van der Waals forces cannot be controlled by an external processing parameter, they are always present; it is fully adequate to make use of a generalized result in praxis.

It was found that van der Waals forces lead to a wetting of the stamp ceiling with gaps of $H - \bar{h} \leq 50$ nm; similarly, de-wetting at the substrate occurs with layers of $\bar{h} \leq 50$ nm. With respect to the guiding chart, it simply means that the upper and the lower boundaries (according to Equations (4) and (5)) have to be shifted by 50 nm downwards and upwards, respectively. This is exemplified in Figure 6 by the lines $\bar{h}_{upp}^{vdW}$ and $\bar{h}_{low}^{vdW}$. The value of 50 nm may be seen as a worst-case limit; it results from a viscosity low enough and an interaction time long enough (namely the size of the parameter P_0); with T-NIL at a viscosity of 10^5 Pas it may be somewhat smaller than 50 nm. As already mentioned in context with Figure 4, at viscosities of $\approx 10^6$ Pas instabilities are not relevant, independent from their source.

5.2.2. Temperature Gradients

The effect of temperature gradients depends on an external control parameter, the size of ΔT, the temperature difference between the surfaces of the stamp, and the substrate with T-NIL. Defect formation is independent from the direction of the temperature gradient [72,74]. As ΔT is not known a priori but system-specific, the respective boundary has to be determined from Equation (9), following the general procedure (vanishing of the right-hand side of the equation). The pressure derivative with temperature gradients reads [45,46]

$$\left(\frac{\partial p_{th}}{\partial h}\right)_{h=\bar{h}} = C_{th}\cdot \left[\kappa_{air}\bar{h} + \kappa_{pol}\left(H - \bar{h}\right)\right]^{-2} \cdot \Delta T, \tag{11}$$

with κ_{air} and κ_{pol} the thermal conductivities of the air gap and the polymer, respectively. The first term of this product is a constant (C_{th}) summarizing thermal material parameters only. The second term combines material parameters and geometries in a way that is typical of temperature gradients. The third term is the external control parameter ΔT. Together with the parameter P_0 already introduced (Equation (9)) the latter defines the process-specific parameter

$$P_{th} = P_0\cdot\Delta T^2. \tag{12}$$

The size of the parameter P_{th} has to be determined by the user. Solutions of Equation (9), regarding the characteristic relationship of Equation (11) can be determined by just considering the value of P_{th}, as $\bar{h}_{th} = \bar{h}(H, P_{th})$. The respective limiting height $\bar{h}_{th}$ is displayed in Figure 7 for values of $10^1 \leq P_{th}/\text{K}^2\text{m}^3\text{N}^{-2} \leq 10^5$. Of course, any polymer height $\bar{h}$ will be smaller than the stamp height H, so that all solutions lie below $\bar{h} = H$ (dashed line in Figure 7).

With respect to the guiding chart (Figure 6), instabilities induced by temperature gradients represent a supplementary boundary limiting the window for defect-free imprint with vanishing residual layer from its top, with small cavity widths. To add this boundary the parameter P_{th} has to be determined from ΔT (maximum or characteristic value), t_p (characteristic time with ΔT), η_p and γ_p, the viscosity and the surface tension of the polymer under processing conditions, here at the respective temperature.

With the value of P_{th} (in units of $\text{K}^2\text{m}^3/\text{N}^2$) the respective maximum height $\bar{h}_{th}$ can be read (and interpolated, if necessary) from Figure 7 and can be drawn as a horizontal line into the guiding chart, Figure 6 (for reasons see Appendix A). The lower $\bar{h}_{th}$, the more the defect-free processing window is clipped at its top. As we found that typically one of the instabilities dominates [46], just the lowest upper limit is effective ($\min\{\bar{h}_{th}, \bar{h}_{upp}^{vdW}\}$). To indicate which limit is dominating, Figure 7 also contains the van der Waals boundary, $\bar{h} = \bar{h}_{upp}^{vdW} = H - 50$ nm (dotted). Thus, only with a process-specific parameter beyond $P_{th} \approx 10^2$ $\text{K}^2\text{m}^3/\text{N}^2$ temperature gradients exceed van der Waals forces and are effective to decrease the defect-free processing window.

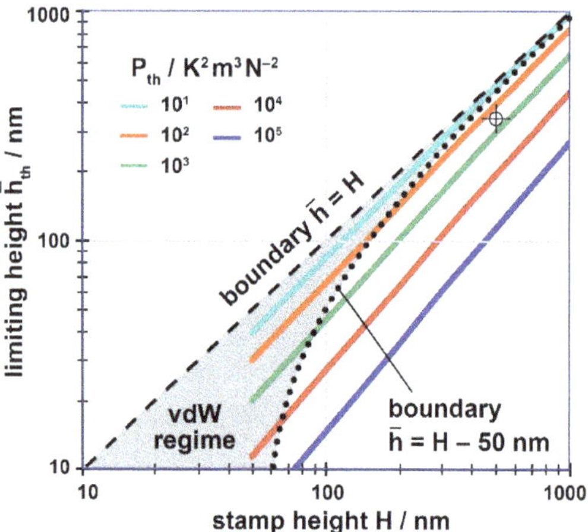

Figure 7. Instabilities due to temperature differences, ΔT. Limiting values of mean polymer height in the cavity, $\bar{h}_{th}$, as a function of the stamp height, H, from which thermal instabilities clip the processing window for defect-free imprint with negligible residual layer. The process-specific parameter P_{th} [K^{2}m^3/N^2] is varied. (Material parameters used [45,72,75]; κ_{air} = 0.034 W/Km, κ_{pol} = 0.16 W/Km, $C_{th} \approx -5 \times 10^{-6}$ (W/K)^3·s/m^4 based on an effective sound velocity of 1850 m/s). The 'cross' refers to the example used for discussion with $P_{th} \approx 300$ K^{2}m^3/N^2, see text.

5.2.3. Electrostatic Forces

The effect of electrostatic forces depends on an external parameter too, the voltage U applied between stamp and substrate (electrically isolated from each other) [70–72,85–87]. The procedure is parallel to the previous one with a temperature gradient. Here the pressure derivative reads [45,46]

$$\left(\frac{\partial p_{el}}{\partial h}\right)_{h=\bar{h}} = C_{el}\cdot \left[\varepsilon_{air}\bar{h} + \varepsilon_{pol}\left(H - \bar{h}\right)\right]^{-3} \cdot U^2, \tag{13}$$

with $\varepsilon_{air} = \varepsilon_0$ and $\varepsilon_{pol} = \varepsilon_0\varepsilon_p$ (ε_0 the dielectric constant) for the air gap and the polymer, respectively. Again, the first term, C_{el}, summarizes material parameters only, now dielectric ones. The second term combines material parameters and geometry parameters in a way that is typical of electrostatic forces (note the exponent differing from Equation (11)) and the third term is the external control parameter, the voltage, entering here as U^2. Similar to the previous case the situation is characterized by a process-specific parameter, now

$$P_{el} = P_0\cdot U^4 \tag{14}$$

The solutions of Equation (9) regarding Equation (13) are plotted in Figure 8; of course, it still holds that $\bar{h} \leq H$. Again, the process-specific parameter P_{el} has to be determined by the user (units now V^{4}m^3/N^2) and a limiting value $\bar{h}_{el}$ for the stamp used (height H) has to be read from Figure 8 and implemented in Figure 6, again a horizontal line at $\bar{h}_{el} = \bar{h}(H, P_{el})$. Only with $P_{el} > 10^6$ V^{4}m^3/N^2 electrostatic instabilities have to be considered in praxis, as evident from Figure 8; with smaller values of P_{el} they are dominated by the van der Waals instabilities always present.

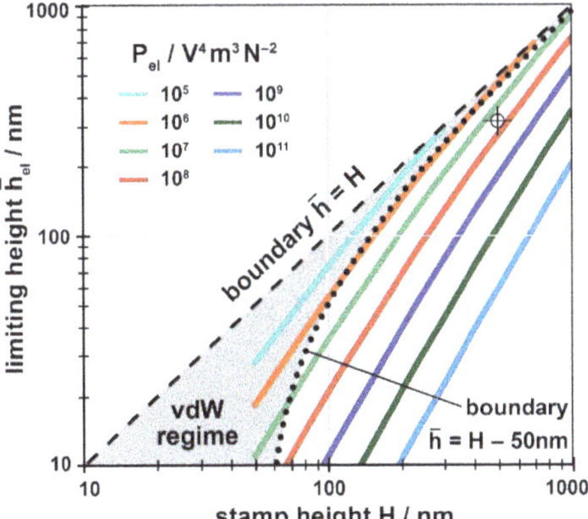

Figure 8. Instabilities due to electrostatic forces (potential difference U). Limiting values of mean polymer height in the cavity, $\bar{h}_{el}$, as a function of the stamp height, H, from which on electrostatic instabilities clip the processing window for defect-free imprint with negligible residual layer. The process-specific parameter P_{el} [V^{4}m^3/N^2] is varied. (Material parameter used; $\varepsilon_p = 3$, $C_{el} \approx -1 \times 10^{-10}$ As/Vm). The 'cross' refers to the example chosen for discussion with $P \approx 3 \times 10^7$ V^{4}m^3/N^2, see text.

6. Working with the Guiding Chart

Though the idea behind the guiding chart is to simplify the use of 'partial cavity filling' with NIL by generalization (in particular the adequate choice of the initial layer thickness h_0), every user has to construct his own guiding chart, being specific for his imprint situation and the imprint system used. The application in mind for nanoimprint as a lithography process has to be included too (lift-off, direct dry etching).

6.1. Construction of a Specific Guiding Chart

For practical use, a specific guiding chart has to consider (i) geometric limitations, (ii) limitations induced by instabilities (with the exception of very high viscosities, $\approx 10^6$ Pas), and (iii) application-specific limitations. To construct the guiding chart the following quantities are required as input parameters.

- Stamp geometries. Height H, cavity width w of interest, duty cycle s/w (see Figure 1).
- Polymer data under processing conditions. Contact angle θ to the stamp, viscosity η_p, surface tension γ_p (estimates may be taken from Table 1).
- System data. Characteristic temperature difference ΔT and/or characteristic voltage U, as well as corresponding interaction time t_p. Please note, ΔT and U refer to the values between the surfaces of substrate and stamp ceiling; these may differ from overall values (e.g., available from data log-files of the imprint system used), depending on the imprint configuration, e.g., thermal/electrical isolation. An estimate may be required.
- Application-related data. Minimum polymer height h^* provided/required for lift-off/etching, with $h^* > 50$ nm (van der Waals limit).

With these data at hand, the construction of the specific guiding chart exploits Figure 5 (alternatively Equation (6)) to find the geometrical limits. To identify the limiting heights induced by thermal or electrostatic instabilities, namely h_{th} and h_{el}, the relationships derived in chapter 5 are required. As already stated there, for a straightforward use, the results are presented in graphical form, Figure 7 referring to thermal instabilities and Figure 8 referring to electrostatic instabilities. To read the limiting height ($\bar{h}_{th}$ or $\bar{h}_{el}$) the value of the respective process-specific parameter, P_{th} or P_{el}, has to be determined from the input parameters, namely $P_{th} = P_0 \cdot \Delta T^2$ (Equation (12)) and/or $P_{el} = P_0 \cdot U^4$ (Equation (14)), with $P_0 = t_p / \eta_p \gamma_p$ (Equation (10)). The specific guiding chart will look similar to Figure 6, however adapted to the characteristic situation of the user.

The construction of the specific guiding chart then proceeds as follows:

1. Draw overall regime. Height $\bar{h} = H$, width about 3–5 times the cavity width w of interest.
2. Add geometric boundaries. Determine m_{upp} and m_{low} from Figure 5 or Equation (6) for the contact angle θ applying. Draw upper boundary $\bar{h}_{upp}$ as a straight line starting at $\bar{h} = H$ at $w = 0$, with a negative slope of m_{upp}; draw $\bar{h}_{low}$ with the positive slope m_{low}, starting from $\bar{h} = 0$ at $w = 0$.
3. Indicate van der Waals limits. Draw $\bar{h}_{upp}^{vdW}$ and $\bar{h}_{low}^{vdW}$ as a parallel to $\bar{h}_{upp}$ and $\bar{h}_{low}$, shifted by 50 nm downwards and upwards, respectively.
4. Add minimum polymer height provided/required for application as a parallel to $\bar{h}_{low}$, shifted upwards by h^*.
5. In case of T-NIL (or electrostatic NIL), add additional upper limit $\bar{h}_{th}$ (or $\bar{h}_{el}$) as a horizontal line; if both apply, take the smaller value. The respective height is read from Figure 7 (or Figure 8) taking the stamp height H and the value of the process-specific parameter applying, P_{th} (or P_{el}).

The processing window (range of values $\bar{h}$ for any cavity width w) available for using partial cavity filling for residual layer-free and defect-free imprint is the regime between the dominating lower limit, $\max\left\{ h*, \bar{h}\,_{low}^{vdW} \right\}$, and the dominating upper limit, $\min\left\{ \bar{h}_{th}, \bar{h}\,_{upp}^{vdW} \right\}$, in analogy to Figure 6.

6.2. Discussion and Conclusions

For a general discussion, we take the guiding chart of our example, Figure 6. The geometries are similar to Figure 4, a stamp of height $H = 500$ nm and a range of cavity widths up to $w = 5$ μm. With thermal and electrostatic instabilities, we address a medium range of the external parameters, $\Delta T \approx 10\ ^\circ$C and $U \approx 10$ V. The respective interaction time t_p for the development of the instabilities matches a specific imprint situation. Typical material parameters with T-NIL refer to Table 1; with UV-NIL the viscosities are in the range of 10^{-1}–10^2 Pas. The choice of the initial layer (see Equation (1), $\bar{h} = h_0(1 + s/w)$) is exemplified by taking cavities of 1 μm width.

From this general discussion and the example with $w = 1$ μm the user may easily draw the conclusions for his own, specific situation, namely an adequate choice of the initial layer height h_0. Furthermore, he may decide which measures are appropriate under his specific conditions/limitations to enlarge the processing window, so that 'partial cavity filling' can be used to imprint with a negligible residual layer, however, without defects.

6.2.1. Lower Limit

At low filling levels, van der Waals forces are the only ones inducing instabilities. In the regime between $\bar{h}\,_{low}^{vdW}$ and $\bar{h}_{low}$ the risk is high that the polymer de-wets on the substrate. With lithography, this will result in mask defects, e.g., holes in the patterned polymer layer. Such structures are definitely unusable for lithography purposes, independent from the strategy followed (lift-off, direct etching). In praxis, there will always be some minimum polymer height h^* required; with lift-off it may be some safety margin; when direct use of the polymeric structures as a mask for dry etching is intended, h^* is determined from the selectivity of the dry etching process applied (as discussed in detail with the geometric processing window, Figure 4). Similar to the purely geometric guiding chart the lower boundary, $\bar{h}\,_{low}^{vdW}$, is mainly of academic interest; it becomes overruled and will be replaced by h^* in praxis, with typically $h^* > \bar{h}\,_{low}^{vdW} = \bar{h}\,_{low} + 50$ nm.

6.2.2. Upper Limit

At high filling levels, $\bar{h}$ is limited by van der Waals forces as well as thermal and electrostatic forces; as in general the lowest limit is dominating, the upper boundary is given by $\min\left\{\bar{h}_{th}, \bar{h}_{el}, \bar{h}\,^{vdW}_{upp}\right\}$ in general. The instabilities defining the upper boundary of the processing window (and thus the choice of the initial layer) depend on the specific technique applied with NIL.

UV-NIL: With UV-NIL neither temperature gradients nor electrostatic fields are present; therefore, just van der Waals forces may cause instabilities. Due to the low viscosity, the van der Waals limit of 50 nm is fully exploited. The upper boundary of the processing window is given by $\bar{h}\,^{vdW}_{upp}$ in Figure 6 ($\bar{h}_{th}$ does not exist), the lower one by h^* (or $\bar{h}\,^{vdW}_{low}$).

In the regime between $\bar{h}\,^{vdW}_{upp}$ and $\bar{h}_{upp}$, there is some risk that polymer bridges to the stamp ceiling develop. As long as their period is small and the polymer height remaining between the bridges is high, this is not critical (whilst h^* is met); an example is given in Figure 2(c1). However, generally there is some risk of de-wetting between the bridges. This is most probable with wide cavities. As already addressed, local bridges lead to a rearrangement of the polymer, enhancing the risk of a local de-wetting of the substrate along the linear cavities, in the third dimension (see Figure 2(a3–5,b3,4)). The rearrangement process is a question of time and viscosity (and thus P_0). To avoid instabilities safely, the regime between $\bar{h}\,^{vdW}_{upp}$ and $\bar{h}_{upp}$ is off-limits with UV-NIL, due to the low viscosity.

As the processing window is not further clipped at its top, it is wide, in particular with small cavities. However, due to the low viscosities and thus a high risk of rearrangement at random flaws of the stamp, the direct neighborhood of the upper boundary should be avoided in praxis.

With 1 µm wide cavities a maximum mean filling height of $\bar{h} \approx 350$ nm should be fully adequate and would provide a good masking height for etching. With a duty cycle of the stamp of $s{:}w$ = 1:1 (2:1) the respective initial height required to provide this filling level amounts to $h_0 \approx 175$ nm (120 nm).

T-NIL: With T-NIL the additional limiting height in the guiding chart is $\bar{h}_{th}$, as indicated in Figure 6, so that the upper boundary of the processing window is represented by $\min\left\{\bar{h}_{th}, \bar{h}\,^{vdW}_{upp}\right\}$. Obviously thermal instabilities are particularly effective with small cavities (high filling levels) and clip the window from its top. Only with T-NIL at untypically low imprint temperatures (viscosities of $\approx 10^6$ Pas) any limitation due to instabilities does not apply (P_0, P_{th} very small). In that case, the upper boundary would simply be $\bar{h}_{upp}$. These conditions would result in the widest defect-free processing window possible.

As a typical example, we determined $P_{th} \approx 300$ K^{2}m^3/N^2 for an imprint with our system at a high temperature, 200 °C, from the values $\Delta T \approx 10$ °C, $t_p \approx 15$ min (during heat-up and cool-down [46]), $\eta_p \approx 10^4$ Pas and $\gamma_p \approx 30$ mJ/m^2. This results in a limiting filling height of $\bar{h}_{th} \approx 320$ nm (see 'cross' in Figure 7). As the viscosity is typically high with T-NIL, random flaws of the stamp are less effective to cause a rearrangement of the polymer within the cavities, so that the remaining processing window may be fully exploited in praxis.

With 1 µm wide cavities, a mean filling height of $\bar{h} \approx 300$ nm should be well suited to avoid instabilities, still a good masking height for etching. With a duty cycle of the stamp of $s{:}w$ = 1:1 (2:1), the respective initial height required to provide this filling level amounts to $h_0 \approx 150$ nm (100 nm). With non-uniformity of the initial layer a somewhat lower value should be chosen.

To illustrate the benefit of the guiding chart the practical examples shown in Figure 2 that were obtained with T-NIL are assigned to the different regions. With a1 we are in the regime of a residual layer, beyond $\bar{h} = H$. With a2 we are within the safe processing window for imprint (Figure 6). The same holds for b1, however depending on the application used, the limit h^* may not be satisfied. With a3–5 and b3,4 we are beyond $\bar{h}_{th}$, the instabilities visible are thermally induced. Lastly, b2 and c4 are beyond $\overline{h}_{upp}$, the meniscus touches the stamp (Figure 4).

el-UV-NIL: With electrostatic nanoimprint, the horizonal line at $\overline{h}_{th}$ is replaced by $\overline{h}_{el}$ in the guiding chart, Figure 6. Therefore, the upper boundary is represented by $\min\left\{\overline{h}_{el}, \overline{h}\,_{upp}^{vdW}\right\}$; again, the window is clipped from its top at small cavity widths (high filling levels).

As a typical example we determined $P_{el} \approx 3 \times 10^7$ V^{4}m^3/N^2, from the values $U = 10$ V, $t_P = 20$ min, $\gamma_p \approx 40$ mJ/m^2 and $\eta_p \approx 10$ Pas. This results in a limit of the filling height of $\overline{h}_{el} \approx 320$ nm (see 'cross' in Figure 8—by chance similar to T-NIL). Due to the low viscosity, again the direct neighborhood of this upper limit should be avoided.

With 1 μm wide cavities a mean filling height of $\overline{h} \approx 270$ nm should be adequate to avoid instabilities. With a duty cycle of the stamp of $s{:}w = 1{:}1$ (2:1) the respective initial height required to provide this filling level amounts to $h_0 \approx 135$ nm (90 nm). With these low initial layer thicknesses non-uniformity may be critical.

el-T-NIL: With both external control parameters, a temperature gradient and a voltage as well, the dominating limit is again the lower one. Accordingly, the upper boundary of the processing window is given by $\min\left\{\overline{h}_{el}, \overline{h}_{th}, \overline{h}\,_{upp}^{vdW}\right\}$; again, the window is clipped from its top at small cavity widths (high filling levels).

As an example, we address a situation met in an earlier investigation, the electrically assisted phase separation of a block-copolymer during T-NIL under 'partial cavity filling' conditions [82]. With this investigation, the voltage drop between substrate and stamp ceiling was about 10 V and the material was treated for 1 h at 170 °C ($\eta \approx 10^5$ Pas, $\gamma_p \approx 33$ mJ/m^2). These values result in a process-specific parameter $P_{el} \approx 10^4$ V^{4}m^3/N^2. With this low value (caused by the high viscosity) electrostatic instabilities will not occur (Figure 8: $P_{el} > 10^6$ V^{4}m^3/N^2 to dominate). Accordingly, our experiments were not affected by electrostatic instabilities (see Figure 2(c2)). The processing window appropriate was the one with T-NIL; accordingly, thermal instabilities were observed at high temperature (see Figure 2(c3)). Here, any minimum height with respect to dry etching was not an issue.

6.2.3. Hidden Control Parameters

When instabilities are considered, the processing window (Figure 6) may seem somewhat constricted, in particular when direct etching with the polymeric mask is intended (h^* high). Nonetheless, the concept of 'partial cavity filling' is powerful to imprint with a negligible residual layer in praxis; yet a well-prepared experiment is asked. The discussion of the processing window available clearly indicated that the upper boundary is the critical one in praxis. Even so, two parameters that are of major impact to widen the processing window may have escaped attention. These somewhat 'hidden' external control parameters are related to the stamp used.

A stamp of adequate height providing a low contact angle widens the processing window substantially (H high, m_{upp} and m_{low} small). In praxis, H may be fixed otherwise. Similarly, choice of the contact angle is restricted as the polymer to be used may be prescribed. However, the contact angle is controlled by the surface properties of the stamp, too. A low surface energy and thus excellent anti-sticking properties are a question of technological diligence with stamp preparation. This emphasizes the impact of a well-controlled and reproducible anti-sticking treatment [63] to limit instabilities.

Furthermore, with small cavities, the upper limit may be further raised when $\overline{h}_{th}$ or $\overline{h}_{el}$ is high. With T-NIL, this is the case when performed under iso-thermal conditions in an imprint system featuring minimum temperature gradients ($\overline{h}_{th}$ high). With el-UV-NIL, a compromise may be required to provide the imprint pressure required, but without being affected by instabilities.

Of course, the (at least approximate) knowledge of the parameters characterizing the imprint system and the imprint material is a prerequisite to fully exploit this concept.

6.2.4. Typical Regimes with NIL

Finally, for practical purposes we illustrate the role of the external control parameters ΔT and U with instabilities. Therefore, regimes prone to instabilities will be indicated with thermal, electrostatic, and electrically assisted NIL. Again, a graphical representation is chosen, shown in Figure 9.

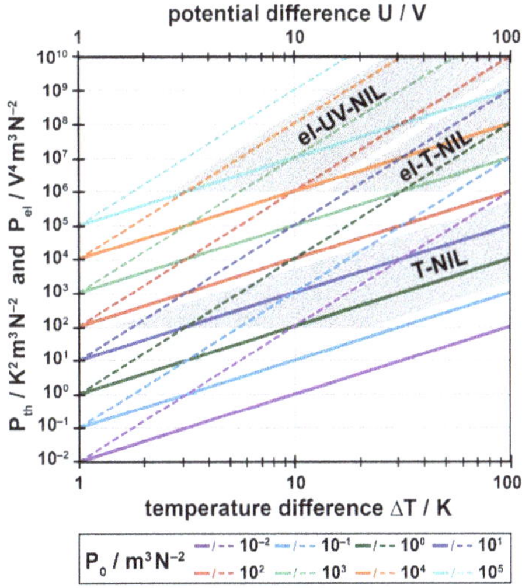

Figure 9. Role of the external control parameters ΔT and U. Process-specific parameters P_{th} (full lines, lower axis) and P_{el} (dashed lines, upper axis), with similar values of the basic processing parameter P_0 (10^{-2}–10^5 m^3/N^2). The marked regions refer to typical parameter ranges with T-NIL, UV-NIL in an electrostatic imprint system (el-UV-NIL) and electrically assisted T-NIL (el-T-NIL), see text.

From Figures 7 and 8, it is evident that the two process-specific parameters, P_{th} and P_{el}, are quite different in size (in parts due to ΔT and U entering by a power of 2 and 4, respectively). Both are compared to each other in Figure 9 to illustrate this difference, with similar values of P_0. In accordance with typical imprint situations, the range of the external control parameters is chosen similarly, $1\,°C \leq \Delta T \leq 100\,°C$ (lower axis) and $1\,V \leq U \leq 100\,V$ (upper axis). The basic processing parameter is varied over seven decades, $10^{-2} \leq P_0 \leq 10^5$.

Of course, P_0 typically is not similar with T-NIL and UV-NIL. As already addressed, the main parameter affecting its size is the viscosity η_p of the imprinted material, which may differ by orders of magnitude. The interaction time t_p may range from 0.5–30 min, depending on the system and the process. The surface tension has the smallest impact. Thus, there is an urgent need to know, at least approximately, the viscosity and the time during which instabilities may develop.

Considering viscosities typical of thermal imprint, P_0 may range from about 0.2 to 50 m^3/N^2. This regime is marked in Figure 9 and assigned as 'T-NIL'. It was considered that thermal instabilities dominate over van der Waals forces only with values of P_{th} beyond $\approx 10^2$ K^{2}m^3/N^2 (see Figure 7). This is the case from about $\Delta T \approx 1\,°C$ on.

Considering viscosities typical of UV-NIL, P_0 may be in the range 70 to 2×10^4 m^3/N^2 for imprint under electrostatic forces, again marked in Figure 9 and assigned as 'el-UV-NIL', dominating from $P_{el} \approx 10^6$ V^{4}m^3/N^2 on. This is the case with voltages of $U \geq 4$ V.

When thermal imprint is combined with electrostatic forces, with or without intention, the range of viscosities (and thus P_0) is similar to T-NIL; this is the regime assigned as 'el-T-NIL'. If electrostatic instabilities are to dominate over van der Waals instabilities, again a value of $P_{el} \approx 10^6$ V^{4}m^3/N^2 has to be exceeded. This happens with more than 10 V only. However, thermal instabilities may also occur and may dominate at low voltages. Thus, both regimes apply in this case, 'el-T-NIL' and 'T-NIL' as well, with electrostatic and thermal instabilities, where the lower limit is the decisive one.

Figure 9 nicely shows that with the different imprint techniques the typical regimes are clearly separated, in terms of their risk of instabilities when varying the external control parameters, ΔT and U. The regions marked in grey indicate the conditions where the undulations are high enough to bridge the gap between mean polymer height and stamp ceiling within typical interaction times. Obviously, T-NIL is most critical, instabilities may develop over the whole range of temperature differences, $1\ ^{\circ}\mathrm{C} < \Delta T < 100\ ^{\circ}\mathrm{C}$. This is different with electrostatic fields. Defects have to be expected from 4V on in an electrostatic UV-NIL system; however, the voltage range allowed without inducing instabilities with a thermal process is substantially higher than with UV-NIL, due to the large difference in the viscosities.

Please note that instabilities due to van der Waals forces are not indicated in Figure 9. They would be present below the respective lower limits of the process-specific parameters, $P_{th} < 10^2\ \mathrm{K^2 m^3/N^2}$ and $P_{el} < 10^6\ \mathrm{V^4 m^3/N^2}$. Similarly, UV-NIL without an electric field cannot be indicated as it becomes limited by van der Waals instabilities only; there is no external control parameter available.

7. Summary

Based on experimental evidence, a guiding chart was developed to facilitate the choice of the initial layer thickness when imprinting periodic structures with a negligible residual layer. The strategy followed is 'partial cavity filling', where a thin layer is printed to its full height so that isolated structures are obtained. When these isolated structures shall serve as a mask for subsequent etching with nanoimprint the lithography technique used, they have to be defect-free. To ensure this, geometric as well as thermodynamic limitations have to be overcome; the latter result from instabilities induced by van der Waals interactions, temperature gradients, or electrostatic forces. Practical use of the concept is encouraged by generalizing the underlying complex physical relationships and by presenting them in simplified form by means of graphs. These graphs can be used when a single, process-specific parameter is at hand. The construction of a tailor-made guiding chart applying to specific imprint situations was demonstrated and the processing window was discussed with T-NIL and UV-NIL, with and without electrostatic forces. Furthermore, measures to enlarge the defect-free processing window were addressed, emphasizing the stamp used and in particular its anti-sticking properties. Examples for an adequate choice of the initial layer thickness based on the respective processing window were given. To widen the applicability, the concept developed in detail with a linear, one-dimensional grating is adapted to a stamp featuring two-dimensional periodic structures.

Author Contributions: Experiments and calculations are done by A.M. A.M. and H.-C.S. contributed equally to conceptualization, methodology, writing and editing of the manuscript. All authors have read and agreed to the published version of the manuscript.

Funding: This research received no external funding.

Acknowledgments: We acknowledge partial funding by the School of Electrical, Information and Media Engineering of the University of Wuppertal. Furthermore, we acknowledge support from the Open Access Publication Fund of the University of Wuppertal.

Conflicts of Interest: The authors declare no conflict of interest.

Appendix A. Periodicity

The periodicity incurring with instabilities differs with their origin [45,46,71,74]. Under conditions typical of nanoimprint, the following holds. Van der Waals instabilities exhibit the widest range of periods possible; periods of a few nanometers are found with narrow gaps but may extend into the millimeter range with wide gaps. With the limit taken above of $\bar{h}_{upp}^{vdW} = (H - 50 \text{ nm})$ the period is around 10 µm. With thermal and electrostatic instabilities, the range is smaller, 1–100 µm. When these instabilities occur together, thermal/electrostatic instabilities dominate with wide gaps (featuring long periods) and van der Waals instabilities dominate with narrow gaps (featuring small periods).

With our cavities we adopt the characteristics of a cavity resonator. When the period of a harmonic oscillation is larger than the dimension of the cavity, this oscillation will not show up (e.g., the case with the examples shown in Figure 2(a3–5), where the oscillation occurs in longitudinal direction but not in transversal direction, other than with Figure 2(b3,c1)). In the course of our simplified handling, we allocate thermal/electrostatic instabilities with a period of >10 µm and van der Waals instabilities with a period <10 µm, down to some 10 nm. Accordingly, van der Waals instabilities 'see' the cavity width and follow the width dependent geometric boundaries, $\bar{h}_{low}(w)$ and $\bar{h}_{upp}(w)$, but thermal/electrostatic instabilities do not; they are indicated in the guiding chart as a horizontal line, independent from the cavity width. As a consequence, the guiding chart introduced (Figure 6) covers cavities of some micrometer in width as an upper limit. With wider cavities the thermal/electrostatic limit would have to follow the cavity size dependence, similar to the van der Waals limit.

Of course, all instabilities will show up along the linear cavities, as addressed in connection with the reconfiguration of the polymer. With a length of linear cavities below $\approx$10 µm, thermal/electrostatic instabilities do not occur (see Appendix B).

Appendix B. 2-Dimensional Gratings

The concept of the guiding chart developed for a linear, one-dimensional grating can be expanded to be used with stamps featuring a two-dimensional grating. The grating then consists of circular or square/rectangular geometries, arranged e.g., in a quadratic or hexagonal grid. Here the duty cycle of the stamp characterizing the material transport into the cavities is given by the respective area ratio A_s/A_w (rather than the ratio of the lateral geometries s/w with one-dimensional gratings). This area ratio reflects the size and distance of the periodic structures, as well as their lateral arrangement, the grid. Again, the guiding chart will be represented as the mean filling height, here according to $\bar{h} = h_0(1 + A_s/A_w)$, over the characteristic size of the cavities, w_c, within a range of interest.

Furthermore, we have to distinguish between 'grid-type' stamps and 'channel-type' stamps. With a 'channel-type' stamp the cavities represent a connected network of channels and the polymer inside these channels behaves similar to a linear grating (compare Figure 6). Therefore, the characteristic width w_c is the minimal distance between the elevated stamp structures. Accordingly, the geometric limitations are given by $\bar{h}_{upp}$ and $\bar{h}_{low}$, where Figure 5 and Equation (6) apply. All other boundaries are as before. Thus, the guiding chart for a 'channel-type' stamp resembles Figure 6; however, the ordinate is now $\bar{h} = h_0(1 + A_s/A_w)$, so that the initial layer height h_0 corresponding to a certain filling level $\bar{h}$ reflects the two-dimensional grating, only. Often the situation is characterized by $A_s/A_w << 1$ so that the initial layer height h_0 is not substantially smaller than the mean filling height $\bar{h}$.

With a 'grid-type' stamp the situation is different as isolated circular or square cavities exist. The characteristic width then is the diameter or the diagonal of the cavities (as already indicated by the experimental result of Figure 2(c4)). In that case, the upper and lower boundaries are again inclined lines in the guiding chart. However, their slope differs from Figure 5 as now the surface of the polymer forming a meniscus is a spherical segment (instead of the circular segment with the 1-dimensional grating). Calculating the respective geometric boundaries proceeds similar to Equations (3)–(5); the respective slopes of the upper and lower boundary are given in Figure A1, again as a function of the contact angle θ. Like before, a linear approximation holds in the range of the contact angles of interest,

$$m_{upp} \approx 0.16 - 1.8 \cdot 10^{-3} \frac{\theta}{\deg} \approx \frac{2}{3} m_{low}. \tag{A1}$$

As obvious from Equation (A1) and Figure A1, now the upper slope is the smaller one. Accordingly, a geometric guiding chart considering the slopes of Figure A1 (or Equation (A1)) applies.

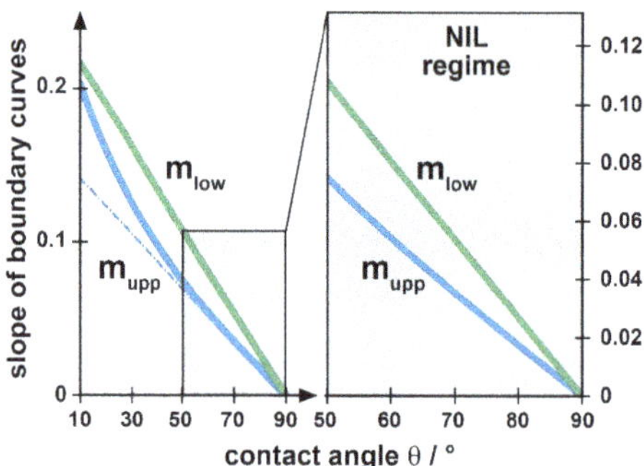

Figure A1. Non-dimensional slopes of the upper (blue) and lower (green) boundary for the formation of a horizontal meniscus in a circular or square/rectangular cavity as a function of the contact angle θ. With high contact angles the linear approximations are adequate (dash-dotted).

To exemplify this situation, a guiding chart (similar to Figure 6) is given in Figure A2, considering a negative stamp with a two-dimensional grating, rectangular cavities (1 μm × 2 μm) at a distance of 1 μm, arranged in a square grid. Then the duty cycle amounts to $A_s/A_w = 4/2 = 2$ and the characteristic cavity width is $w_c \approx 2.2$ μm. Due to the limitation of the cavity in two dimensions, only van der Waals instabilities occur (see Appendix A).

With our example of Figure A2, a mean filling height of $\bar{h} \approx 300$ nm could be adequate to avoid defects. Due to the duty cycle already high of $A_s/A_w = 2$ this requires an initial layer as low as 100 nm. This example clearly shows that despite the wide-looking processing window the realization of a defect-free imprint with isolated cavities may be hampered by preparing a very thin residual layer of high uniformity. Of course, this is due to the positive nature of the stamp. Generally, a similar issue exists with linear cavities, however typical duty cycles are not as high as with two-dimensional gratings.

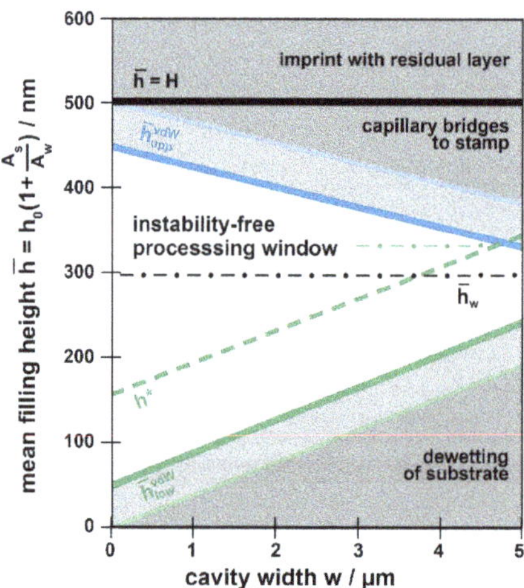

Figure A2. Example of a modified guiding chart for defect-free imprint with negligible residual layer similar to Figure 6, applying to a one-dimensional grating with isolated cavities. The processing window (bright) is limited by van der Waals instabilities, only.

References

1. Chou, S.Y.; Krauss, P.R.; Renstrom, P.J. Imprint of sub-25 nm vias and trenches in polymers. *Appl. Phys. Lett.* **1995**, *67*, 3114–3116. [CrossRef]
2. Haisma, J. Mold-assisted nanolithography: A process for reliable pattern replication. *J. Vac. Sci. Technol. B Microelectron. Nanometer Struct.* **1996**, *14*, 4124. [CrossRef]
3. Colburn, M.; Johnson, S.C.; Stewart, M.D.; Damle, S.; Bailey, T.C.; Choi, B.; Wedlake, M.; Michaelson, T.B.; Sreenivasan, S.V.; Ekerdt, J.G.; et al. Step and flash imprint lithography: A new approach to high-resolution patterning. *Emerg. Lithogr. Technol. III* **1999**, *3676*, 379–389. [CrossRef]
4. Hensel, R.; Finn, A.; Helbig, R.; Braun, H.-G.; Neinhuis, C.; Fischer, W.-J.; Werner, C. Biologically Inspired Omniphobic Surfaces by Reverse Imprint Lithography. *Adv. Mater.* **2014**, *26*, 2029–2033. [CrossRef]
5. Meudt, M.; Bogiadzi, C.; Wrobel, K.; Görrn, P. Hybrid Photonic—Plasmonic Bound States in Continuum for Enhanced Light Manipulation. *Adv. Opt. Mater.* **2020**, *8*. [CrossRef]
6. Haslinger, M.J.; Moharana, A.R.; Mühlberger, M. Antireflective moth-eye structures on curved surfaces fabricated by nanoimprint lithography. *Proc. SPIE* **2019**, *11177*, 111770K. [CrossRef]
7. Kirchner, R.; Finn, A.; Landgraf, R.; Nueske, L.; Teng, L.; Vogler, M.; Fischer, W.-J. Direct UV-Imprinting of Hybrid-Polymer Photonic Microring Resonators and Their Characterization. *J. Light. Technol.* **2014**, *32*, 1674–1681. [CrossRef]
8. Jaszewski, R.; Schift, H.; Gobrecht, J.; Smith, P. Hot embossing in polymers as a direct way to pattern resist. *Microelectron. Eng.* **1998**, *41–42*, 575–578. [CrossRef]
9. Guo, L.J. Recent progress in nanoimprint technology and its applications. *J. Phys. D Appl. Phys.* **2004**, *37*, R123–R141. [CrossRef]
10. Cross, G.L.W. The production of nanostructures by mechanical forming. *J. Phys. D Appl. Phys.* **2006**, *39*, R363–R386. [CrossRef]
11. Guo, L.J. Nanoimprint Lithography: Methods and Material Requirements. *Adv. Mater.* **2007**, *19*, 495–513. [CrossRef]
12. Schift, H. Nanoimprint lithography: An old story in modern times? A review. *J. Vac. Sci. Technol. B Microelectron. Nanometer Struct.* **2008**, *26*, 458. [CrossRef]
13. Malloy, M. Technology review and assessment of nanoimprint lithography for semiconductor and patterned media manufacturing. *J. Micro Nanolithogr. MEMS MOEMS* **2011**, *10*, 032001. [CrossRef]
14. Kooy, N.; Mohamed, K.; Pin, L.T.; Guan, O.S. A review of roll-to-roll nanoimprint lithography. *Nanoscale Res. Lett.* **2014**, *9*, 320. [CrossRef] [PubMed]
15. Schift, H. Nanoimprint lithography: 2D or not 2D? A review. *Appl. Phys. A* **2015**, *121*, 415–435. [CrossRef]
16. Zhang, C.; Subbaraman, H.; Li, Q.; Pan, Z.; Ok, J.G.; Ling, T.; Chung, C.-J.; Zhang, X.; Lin, X.; Chen, R.T.; et al. Printed photonic elements: Nanoimprinting and beyond. *J. Mater. Chem. C* **2016**, *4*, 5133–5153. [CrossRef]
17. Verschuuren, M.A.; Megens, M.; Ni, Y.; Van Sprang, H.; Polman, A. Large area nanoimprint by substrate conformal imprint lithography (SCIL). *Adv. Opt. Technol.* **2017**, *6*, 243–264. [CrossRef]
18. Resnick, D.J.; Choi, J. A review of nanoimprint lithography for high-volume semiconductor device manufacturing. *Adv. Opt. Technol.* **2017**, *6*, 229–241. [CrossRef]
19. Shao, J.; Chen, X.; Li, X.; Tian, H.; Wang, C.; Lu, B. Nanoimprint lithography for the manufacturing of flexible electronics. *Sci. China Ser. E Technol. Sci.* **2019**, *62*, 175–198. [CrossRef]
20. Heyderman, L.J.; Schift, H.; David, C.; Gobrecht, J.; Schweizer, T. Flow behaviour of thin polymer films used for hot embossing lithography. *Microelectron. Eng.* **2000**, *54*, 229–245. [CrossRef]
21. Schift, H.; Heyderman, L.J.; Der Maur, M.A.; Gobrecht, J. Pattern formation in hot embossing of thin polymer films. *Nanotechnology* **2001**, *12*, 173–177. [CrossRef]
22. Sakamoto, J.; Fujikawa, N.; Nishikura, N.; Kawata, H.; Yasuda, M.; Hirai, Y. High aspect ratio fine pattern transfer using a novel mold by nanoimprint lithography. *J. Vac. Sci. Technol. B* **2011**, *29*, 06FC15. [CrossRef]
23. Gourgon, C.; Perret, C.; Tallal, J.; Lazzarino, F.; Landis, S.; Joubert, O.; Pelzer, R. Uniformity across 200 mm silicon wafers printed by nanoimprint lithography. *J. Phys. D Appl. Phys.* **2004**, *38*, 70–73. [CrossRef]
24. Cheng, X.; Guo, L.J.; Fu, P.-F. Room-Temperature, Low-Pressure Nanoimprinting Based on Cationic Photopolymerization of Novel Epoxysilicone Monomers. *Adv. Mater.* **2005**, *17*, 1419–1424. [CrossRef]
25. Vratzov, B.; Fuchs, A.; Lemme, M.; Henschel, W.; Kurz, H. Large scale ultraviolet-based nanoimprint lithography. *J. Vac. Sci. Technol. B Microelectron. Nanometer Struct.* **2003**, *21*, 2760. [CrossRef]
26. Schmitt, H.; Frey, L.; Ryssel, H.; Rommel, M.; Lehrer, C. UV nanoimprint materials: Surface energies, residual layers, and imprint quality. *J. Vac. Sci. Technol. B Microelectron. Nanometer Struct.* **2007**, *25*, 785–790. [CrossRef]
27. Kirchner, R.; Nüske, L.; Finn, A.; Lu, B.; Fischer, W.-J. Stamp-and-repeat UV-imprinting of spin-coated films: Pre-exposure and imprint defects. *Microelectron. Eng.* **2012**, *97*, 117–121. [CrossRef]
28. Chen, Y.; Carcenac, F.; Ecoffet, C.; Lougnot, D.; Launois, H. Mold-assisted near-field optical lithography. *Microelectron. Eng.* **1999**, *46*, 69–72. [CrossRef]
29. Auner, C.; Palfinger, U.; Gold, H.; Kraxner, J.; Haase, A.; Haber, T.; Sezen, M.; Grogger, W.; Jakopic, G.; Krenn, J.; et al. Residue-free room temperature UV-nanoimprinting of submicron organic thin film transistors. *Org. Electron.* **2009**, *10*, 1466–1472. [CrossRef]
30. Kolli, V.; Woidt, C.; Hillmer, H. Residual-layer-free 3D nanoimprint using hybrid soft templates. *Microelectron. Eng.* **2016**, *149*, 159–165. [CrossRef]

31. Dhima, K.; Steinberg, C.; Mayer, A.; Wang, S.; Papenheim, M.; Scheer, H.-C. Residual layer lithography. *Microelectron. Eng.* **2014**, *123*, 84–88. [CrossRef]
32. Cheng, X.; Guo, L.J. One-step lithography for various size patterns with a hybrid mask-mold. *Microelectron. Eng.* **2004**, *71*, 288–293. [CrossRef]
33. Gourgon, C.; Perret, C.; Micouin, G.; Lazzarino, F.; Tortai, J.H.; Joubert, O.; Grolier, J.-P.E. Influence of pattern density in nanoimprint lithography. *J. Vac. Sci. Technol. B Microelectron. Nanometer Struct.* **2003**, *21*, 98. [CrossRef]
34. Schulz, H.; Wissen, M.; Scheer, H.-C. Local mass transport and its effect on global pattern replication during hot embossing. *Microelectron. Eng.* **2003**, *67–68*, 657–663. [CrossRef]
35. Lazzarino, F.; Gourgon, C.; Schiavone, P.; Perret, C. Mold deformation in nanoimprint lithography. *J. Vac. Sci. Technol. B Microelectron. Nanometer Struct.* **2004**, *22*, 3318. [CrossRef]
36. Merino, S.; Retolaza, A.; Juarros, A.; Schift, H. The influence of stamp deformation on residual layer homogeneity in thermal nanoimprint lithography. *Microelectron. Eng.* **2008**, *85*, 1892–1896. [CrossRef]
37. Chaix, N.; Gourgon, C.; Perret, C.; Landis, S.; Leveder, T. Nanoimprint lithography processes on 200 mm Si wafer for optical application: Residual thickness etching anisotropy. *J. Vac. Sci. Technol. B Microelectron. Nanometer Struct.* **2007**, *25*, 2346. [CrossRef]
38. Scheer, H.-C. Pattern definition by nanoimprint. *Photonics Eur.* **2012**, *8428*, 842802. [CrossRef]
39. Suzuki, K.; Youn, S.-W.; Wang, Q.; Hiroshima, H.; Nishioka, Y. Fabrication Processes for Capacity-Equalized Mold with Fine Patterns. *Jpn. J. Appl. Phys.* **2011**, *50*, 06GK04. [CrossRef]
40. Suh, K.; Lee, H. Capillary Force Lithography: Large-Area Patterning, Self-Organization, and Anisotropic Dewetting. *Adv. Funct. Mater.* **2002**, *12*, 405–413. [CrossRef]
41. Li, X.; Ding, Y.; Shao, J.; Tian, H.; Liu, H. Fabrication of Microlens Arrays with Well-controlled Curvature by Liquid Trapping and Electrohydrodynamic Deformation in Microholes. *Adv. Mater.* **2012**, *24*, OP165–OP169. [CrossRef]
42. Bogdanski, N.; Wissen, M.; Möllenbeck, S.; Scheer, H.C. Thermal imprint with negligibly low residual layer. *J. Vac. Sci. Technol. B Microelectron. Nanometer Struct.* **2006**, *24*, 2998. [CrossRef]
43. Lee, H.-J.; Ro, H.W.; Soles, C.L.; Jones, R.L.; Lin, E.K.; Wu, W.-L.; Hines, D.R. Effect of initial resist thickness on residual layer thickness of nanoimprinted structures. *J. Vac. Sci. Technol. B Microelectron. Nanometer Struct.* **2005**, *23*, 3023. [CrossRef]
44. Bogdanski, N.; Wissen, M.; Möllenbeck, S.; Scheer, H.-C. Challenges of residual layer minimisation in thermal nanoimprint lithography. *Proc. SPIE* **2007**, *6533*, 65330Q. [CrossRef]
45. Mayer, A. Self-Assembled Structures in Thermal Nanoimprint. Ph.D. Thesis, University of Wuppertal, Neopubli GmbH, Berlin, Germany, 2018. ISBN 978-3-7467-5954-8.
46. Mayer, A.; Dhima, K.; Wang, S.; Steinberg, C.; Papenheim, M.; Scheer, H.-C. The underestimated impact of instabilities with nanoimprint. *Appl. Phys. A* **2015**, *121*, 405–414. [CrossRef]
47. Chaix, N.; Gourgon, C.; Landis, S.; Perret, C.; Fink, M.; Reuther, F.; Mecerreyes, D. Influence of the molecular weight and imprint conditions on the formation of capillary bridges in nanoimprint lithography. *Nanotechnology* **2006**, *17*, 4082–4087. [CrossRef] [PubMed]
48. Suh, K.Y.; Lee, H.H. Self-Organized Polymeric Microstructures. *Adv. Mater.* **2002**, *14*, 346–351. [CrossRef]
49. Scheer, H.-C.; Schulz, H. A contribution to the flow behaviour of thin polymer films during hot embossing lithography. *Microelectron. Eng.* **2001**, *56*, 311–332. [CrossRef]
50. Chou, S.Y.; Zhuang, L.; Guo, L. Lithographically induced self-construction of polymer microstructures for resistless patterning. *Appl. Phys. Lett.* **1999**, *75*, 1004–1006. [CrossRef]
51. Chou, S.Y.; Zhuang, L. Lithographically induced self-assembly of periodic polymer micropillar arrays. *J. Vac. Sci. Technol. B Microelectron. Nanometer Struct.* **1999**, *17*, 3197. [CrossRef]
52. Pease, L.F.; Russel, W.B. Electrostatically induced submicron patterning of thin perfect and leaky dielectric films: A generalized linear stability analysis. *J. Chem. Phys.* **2003**, *118*, 3790. [CrossRef]
53. Wu, N.; Pease, L.F.; Russel, W.B. Toward Large-Scale Alignment of Electrohydrodynamic Patterning of Thin Polymer Films. *Adv. Funct. Mater.* **2006**, *16*, 1992–1999. [CrossRef]
54. Wu, N.; Russel, W.B. Micro- and nano-patterns created via electrohydrodynamic instabilities. *Nano Today* **2009**, *4*, 180–192. [CrossRef]
55. Yoon, H.; Lee, S.H.; Sung, S.H.; Suh, K.Y.; Char, K. Mold Design Rules for Residual Layer-Free Patterning in Thermal Imprint Lithography. *Langmuir* **2011**, *27*, 7944–7948. [CrossRef]
56. Yoon, H.; Choi, M.K.; Suh, K.Y.; Char, K. Self-modulating polymer resist patterns in pressure-assisted capillary force lithography. *J. Colloid Interface Sci.* **2010**, *346*, 476–482. [CrossRef]
57. Scheer, H.C.; Bogdanski, N.; Wissen, M.; Möllenbeck, S. Impact of glass temperature for thermal nanoimprint. *J. Vac. Sci. Technol. B Microelectron. Nanometer Struct.* **2007**, *25*, 2392. [CrossRef]
58. Schulz, H.; Wissen, M.; Bogdanski, N.; Scheer, H.-C.; Mattes, K.; Friedrich, C. Choice of the molecular weight of an imprint polymer for hot embossing lithography. *Microelectron. Eng.* **2005**, *78-79*, 625–632. [CrossRef]
59. Bogdanski, N.; Wissen, M.; Möllenbeck, S.; Scheer, H.C. Thermal imprint into thin polymer layers below the critical molecular weight. *J. Vac. Sci. Technol. B Microelectron. Nanometer Struct.* **2009**, *27*, 1191. [CrossRef]
60. Scheer, H.-C.; Bogdanski, N.; Wissen, M.; Möllenbeck, S. Imprintability of polymers for thermal nanoimprint. *Microelectron. Eng.* **2008**, *85*, 890–896. [CrossRef]

61. Bogdanski, N.; Möllenbeck, S.; Scheer, H.-C. Contact angles in a thermal imprint process. *J. Vac. Sci. Technol. B Microelectron. Nanometer Struct.* **2008**, *26*, 2416–2420. [CrossRef]
62. Papenheim, M.; Mayer, A.; Wang, S.; Steinberg, C.; Scheer, H.-C. Flat and highly flexible composite stamps for nanoimprint, their preparation and their limits. *J. Vac. Sci. Technol. B* **2016**, *34*, 06K406. [CrossRef]
63. Steinberg, C.; Dhima, K.; Blenskens, D.; Mayer, A.; Wang, S.; Papenheim, M.; Scheer, H.-C.; Zajadacz, J.; Zimmer, K. A scalable anti-sticking layer process via controlled evaporation. *Microelectron. Eng.* **2014**, *123*, 4–8. [CrossRef]
64. Mayer, A.; Moellenbeck, S.; Dhima, K.; Wang, S.; Scheer, H.-C. Mechanical characterization of a piezo-operated thermal imprint system. *J. Vac. Sci. Technol. B* **2011**, *29*, 6. [CrossRef]
65. Mayer, A.; Dhima, K.; Möllenbeck, S.; Wang, S.; Scheer, H.-C. A novel tool for frequency assisted thermal nanoimprint (T-NIL). *Proc. SPIE* **2012**, *8352*, 83520N. [CrossRef]
66. McMackin, I.; Schumaker, P.; Babbs, D.; Choi, J.; Collison, W.; Sreenivasan, S.V.; Schumaker, N.E.; Watts, M.P.C.; Voisin, R.D. Design and performance of a step and repeat imprinting machine. *Microlithography* **2003**, *5037*, 178–186. [CrossRef]
67. Dhima, K. Hybrid lithography: The combination of T-NIL & UV-L. Ph.D. Thesis, University of Wuppertal, Berlin, Germany, 2014. ISBN 978-3-86247-502-5.
68. Vrij, A. Possible mechanism for the spontaneous rupture of thin, free liquid films. *Discuss. Faraday Soc.* **1966**, *42*, 23–33. [CrossRef]
69. Landis, S.; Chaix, N.; Hermelin, D.; Leveder, T.; Gourgon, C. Investigation of capillary bridges growth in NIL process. *Microelectron. Eng.* **2007**, *84*, 940–944. [CrossRef]
70. Pease, L.F.; Russel, W.B.; Iii, L.F.P. Linear stability analysis of thin leaky dielectric films subjected to electric fields. *J. Non Newton. Fluid Mech.* **2002**, *102*, 233–250. [CrossRef]
71. Schäffer, E.; Thurn-Albrecht, T.; Russell, T.P.; Steiner, U. Electrohydrodynamic instabilities in polymer films. *EPL Europhys. Lett.* **2001**, *53*, 518–524. [CrossRef]
72. Schäffer, E. *Instabilities in Thin Polymer Films: Structure Formation and Pattern Transfer*; University of Konstanz: Konstanz, Germany, 2001.
73. Faber, T.E. *Fluid Dynamics for Physicists*, 1st ed.; Cambridge University Press: Cambridge, UK, 1995; ISBN 0 521 41943 3.
74. Schäffer, E.; Harkema, S.; Roerdink, M.; Blossey, R.; Steiner, U. Thermomechanical Lithography: Pattern Replication Using a Temperature Gradient Driven Instability. *Adv. Mater.* **2003**, *15*, 514–517. [CrossRef]
75. Schäffer, E.; Harkema, S.; Roerdink, M.; Blossey, R.; Steiner, U. Morphological Instability of a Confined Polymer Film in a Thermal Gradient. *Macromolecules* **2003**, *36*, 1645–1655. [CrossRef]
76. Israelachvili, J.N. The calculation of van der Waals dispersion forces between macroscopic bodies. *Proc. R. Soc. London. Ser. A Math. Phys. Sci.* **1972**, *331*, 39–55. [CrossRef]
77. Tormen, M.; Malureanu, R.; Pedersen, R.H.; Lorenzen, L.V.; Rasmussen, K.H.; Lüscher, C.J.; Kristensen, A.; Hansen, O. Fast thermal nanoimprint lithography by a stamp with integrated heater. *Microelectron. Eng.* **2008**, *85*, 1229–1232. [CrossRef]
78. Pedersen, R.H.; Hansen, O.; Kristensen, A. A compact system for large-area thermal nanoimprint lithography using smart stamps. *J. MicroMech. MicroEng.* **2008**, *18*, 055018. [CrossRef]
79. Fu, X.; Chen, Q.; Chen, X.; Zhang, L.; Yang, A.; Cui, Y.; Yuan, C.; Ge, H. A Rapid Thermal Nanoimprint Apparatus through Induction Heating of Nickel Mold. *Micromachines* **2019**, *10*, 334. [CrossRef]
80. Liang, X.; Zhang, W.; Li, M.; Xia, Q.; Wu, W.; Ge, H.; Huang, X.; Chou, S.Y. Electrostatic Force-Assisted Nanoimprint Lithography (EFAN). *Nano Lett.* **2005**, *5*, 527–530. [CrossRef] [PubMed]
81. Thurn-Albrecht, T.; DeRouchey, J.; Russell, T.P.; Jaeger, H.M. Overcoming Interfacial Interactions with Electric Fields. *Macromolecules* **2000**, *33*, 3250–3253. [CrossRef]
82. Mayer, A.; Ai, W.; Rond, J.; Staabs, J.; Steinberg, C.; Papenheim, M.; Scheer, H.-C.; Tormen, M.; Cian, A.; Zajadacz, J.; et al. Electrically-assisted nanoimprint of block copolymers. *J. Vac. Sci. Technol. B* **2019**, *37*, 011601. [CrossRef]
83. Mayer, A.; Dhima, K.; Wang, S.; Scheer, H.-C.; Möllenbeck, S.; Sakamoto, J.; Kawata, H.; Hirai, Y. Study of defect mechanisms in partly filled stamp cavities for thermal nanoimprint control. *J. Vac. Sci. Technol. B* **2012**, *30*, 06FB03. [CrossRef]
84. Israrelachvili, J.N. *Intermolecular and Surface Forces*, 3rd ed.; Elsevier Inc.: Burlington, VT, USA, 2011; ISBN 978-0-12-375182-9.
85. Schäffer, E.; Thurn-Albrecht, T.; Russell, T.P.; Steiner, U. Electrically induced structure formation and pattern transfer. *Nat. Cell Biol.* **2000**, *403*, 874–877. [CrossRef]
86. Lee, S.; Jung, S.; Jang, A.-R.; Hwang, J.; Shin, H.S.; Lee, J.; Kang, D.J. An innovative scheme for sub-50 nm patterning via electrohydrodynamic lithography. *Nanoscale* **2017**, *9*, 11881–11887. [CrossRef] [PubMed]
87. Shankar, V.; Sharma, A. Instability of the interface between thin fluid films subjected to electric fields. *J. Colloid Interface Sci.* **2004**, *274*, 294–308. [CrossRef] [PubMed]

Article

Transfer Durability of Line-Patterned Replica Mold Made of High-Hardness UV-Curable Resin

Tetsuma Marumo [1], Shin Hiwasa [2] and Jun Taniguchi [1],*

[1]   Department of Applied Electronics, Tokyo University of Science, 6-3-1, Niijyuku, Katsushika-ku, Tokyo 125-8585, Japan; 8115133@alumni.tus.ac.jp

[2]   Autex Inc., 16-5 Tomihisa, Shinjuku, Tokyo 162-0067, Japan; s_hiwasa@autex-inc.co.jp

*   Correspondence: junt@te.noda.tus.ac.jp; Tel.: +81-3-5876-1440

Received: 31 August 2020; Accepted: 28 September 2020; Published: 1 October 2020

Abstract: Ultraviolet nanoimprint lithography (UV-NIL) requires high durability of the mold for the mass production of nanostructures. To evaluate the durability of a line-patterned replica mold made of high-hardness UV curable resin, repetitive transfer and contact angle measurements of the replica mold were carried out. In the line patterns, as the contact angle decreases due to repeated transfer, capillary action occurs, and water flows along them. Therefore, it can be said that a mold with a line pattern exhibits an anisotropic contact angle because these values vary depending on the direction of the contact angle measurement. Subsequently, these anisotropic characteristics were investigated. It was determined that it was possible to predict the lifetime of line-and-space molds over repeated transfers. As the transcription was repeated, the contact angle along the line patterns decreased significantly before becoming constant. Moreover, the contact angle across the line pattern decreased slowly while maintaining a high contact angle with respect to the contact angle along the line pattern. The contact angle then decreased linearly from approximately 90°. The mold was found to be macroscopically defect when the values of the contact angle along the line pattern and the contact angle across the line pattern were close. Predicting the mold's lifetime could potentially lead to a shortened durability evaluation time and the avoidance of pattern defects.

Keywords: ultraviolet nanoimprint lithography; durability; anisotropy; contact angle; line and space

---

1. Introduction

Ultraviolet nanoimprint lithography (UV-NIL) can be used to fabricate nanoscale patterns without many process steps [1]. It allows numerous nanoscale items to be manufactured at low cost and in high numbers [2,3]. The master mold is usually an expensive material, such as Si, which takes time and labor to process [4,5]. Therefore, it is necessary to prevent damage to the master mold due to resin adhesion. Coating the master mold with a release agent on the surface can prevent such damage [6].

However, repeated transfer degrades the mold release agent and eventually damages the master mold [6–9]. Methods for avoiding damage to the master mold include using a replica mold [10–12] and predicting the master mold's lifetime. To the best of our knowledge, there are no studies on the prediction of mold lifetime; therefore, the durability of replica molds is often evaluated alternatively. Specifically, the durability can be evaluated by measuring the water contact angle, which changes with the water height because of the surface free energy of the mold [13,14]. In the line pattern, water flows along the lines as water height decreases; therefore, the line pattern exhibits contact angle anisotropy [15].

In the current study, the durability of line patterns with anisotropic contact angles was evaluated, and the anisotropy characteristics of the contact angle were investigated. A replica mold for evaluating durability was fabricated using a release-agent-free high-hardness resin.

As illustrated in Figure 1, the contact angle in the x direction (θ_x), which is along the line pattern, and the contact angle in the y direction (θ_y), which is across the line pattern, were measured. The anisotropy was then evaluated by comparing the measured θ_x and θ_y.

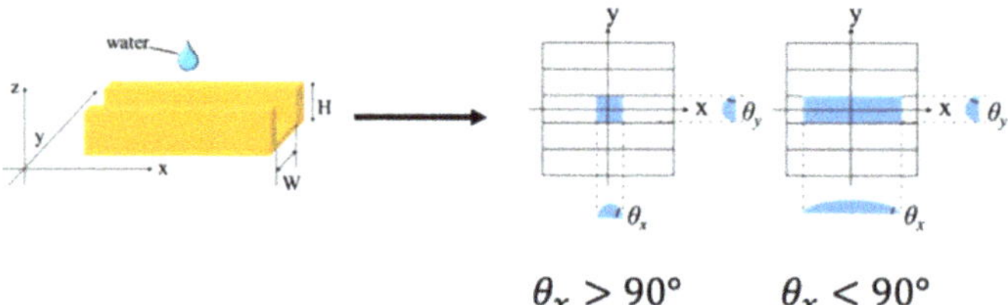

Figure 1. Contact angle measurements for the line-patterned replica mold.

2. Theory

Figure 1 depicts a schematic of the water flow on the line-patterned replica mold. The penetration distance of water in the capillary tube can be expressed as follows: The droplet position at the time of the contact angle measurement is considered as the origin position. The mold line direction, line width direction, and the height are represented by x, y, and z, respectively. First, the flow of water on a plane is expressed by an equation. The flow velocity is set to zero as the boundary condition between the fluid and the wall. The time term is omitted because the system is under steady flow conditions [16], and the external force term is omitted in the absence of any external force. When the fluid flows with a constant pressure gradient, the flow velocity of water is expressed by Equation (1) [17].

$$u = \frac{1}{2\mu}\frac{\partial P}{\partial x}\left(y^2 - Wy\right) = \frac{1}{2\mu}\frac{\Delta P}{l}\left(y^2 - Wy\right) \tag{1}$$

where, u is the flow velocity, μ is the viscosity coefficient, ΔP is the pressure difference between the inside and outside of the liquid interface, W is the width of the second replica mold, l is the penetration distance, and H is the height. To obtain the water penetration distance, the flow rate, Q, given in Equation (2) is obtained from Equation (1). The flow rate of the line-patterned structure is obtained by integrating the equation of the flow velocity by the width and multiplying the equation by the height. The resulting equation is as follows:

$$Q = H\int_0^W u\,dy = -\frac{H}{12\mu}\frac{\partial P}{\partial x}W^3. \tag{2}$$

Q can also be expressed by Equation (3).

$$Q = \frac{dP}{dt} = \frac{WHdl}{dt} \tag{3}$$

Later, Equation (4) is obtained from Equations (2) and (3) and represents the water penetration distance:

$$-\frac{H}{12\mu}\frac{\Delta P}{l}W^3 = \frac{WHdl}{dt}$$

$$ldl = -\frac{\Delta P}{12\mu}W^2 dt$$

$$l = \sqrt{\left|\frac{\Delta P}{6\mu}W^2 t\right|} \tag{4}$$

where, t is the water penetration time and l is the water penetration distance. Furthermore, the pressure difference ΔP can be obtained from Equation (5) [17]. However, in the line-and-space (L&S) pattern, the top surface was open; consequently, the value of $\cos\theta$ for the top surface was zero.

$$\Delta P = \frac{(2H + W)\delta\cos\theta}{WH} \tag{5}$$

where, δ is the surface tension of water and θ is the contact angle. The pressure is positive or negative depending on the value of $\cos\theta$. Therefore, ΔP is negative when θ is greater than 90° and positive when θ is less than 90°. Moreover, the direction of the capillary force varies depending on whether θ is ≥90° or <90° [18].

3. Materials and Methods

The first replica mold was fabricated [19] as follows: UV-curable resin, PAK-01-CL (Toyo Gosei Co., Ltd., Tokyo, Japan) was dropped on a Si master mold (Toppan Co., Ltd., Tokyo, Japan). PAK-01-CL is an acrylic base resin and a radical polymerization system. For this purpose, Si master molds with line widths of 100 nm and 200 nm were used. The periodicity of the first master mold was 100 nm line width and 100 nm spacing, whereas that of the second was 200 nm line width and 200 nm spacing. Subsequently, the PAK-01-CL was filled on the Si master mold by pressing the film (Cosmoshine A4300; Toyobo Co., Ltd., Osaka, Japan). The PAK-01-CL was then cured by UV irradiation at a dose of 120 mJ/cm^2 using a UV lamp (Aicure UP50 (Panasonic Co., Ltd., Osaka, Japan), wavelength: 365 nm). The mold fabricated by release was labeled as the first replica mold.

After fabricating the first replica mold, mold release treatment was performed on its surface for improving the mold release. Specifically, 10-nm-thick Pt was deposited on the mold surface, and the mold was then immersed in Optool DSX 0.1% (Daikin Co., Ltd., Osaka, Japan) for 24 h. Later, heating was performed at 85 °C for 30 min to make the fluorinated hydrocarbon chains on the mold surface face upward [20]. The mold was then rinsed for 1 min with Optool HD-TH (Daikin Co., Ltd., Osaka, Japan). Finally, a second replica mold was fabricated from the first replica mold through UV-NIL. Specifically, a release-agent-free high-hardness resin (PARQIT OEX-066-X1-3; Autex Co., Ltd., Tokyo, Japan) was used for its fabrication [19]. This resin exhibits a viscosity in the range of 60–150 mPa·s at 23 °C, a pencil hardness of 9H, and a Young's modulus of 1305 MPa. Because this resin was a release-agent-free resin, there was no need to perform release treatment on it [21]. A UV dose of 50 J/cm^2 was provided by a UV lamp (Aicure UP50 (Panasonic Co.), wavelength: 365 nm). During UV irradiation, the mold was heated at 80 °C to promote curing [19]. After mold release, it was heated at 100 °C for 30 min to improve its mold release ability. These replica molds were prepared with 100 nm line-and-space (L&S) and 200 nm L&S patterns. The first and second replica molds were formed on a polyester film (COSMOSHINE 4300, Toyobo Co., Ltd., Osaka, Japan). The reason behind creating a replica mold of PAK-01-CL was to protect the silicon master mold. This was because PARQIT OEX-0066-X1-3 sometimes partially adhered to the silicon master mold surface even though the surface is released. In contrast, because PAK-01-CL exhibited decent release properties for the silicon master mold, we used the PAK-01-CL replica mold to fabricate the PARQIT OEX-0066-X1-3 replica mold. Figure 2 depicts the scanning electron microscopy (SEM) images of the first and second replica molds.

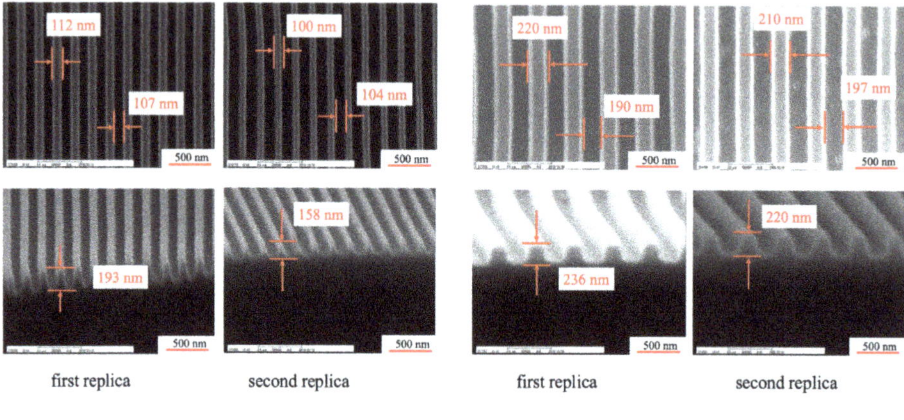

<table>
<tr><td>first replica</td><td>second replica</td><td>first replica</td><td>second replica</td></tr>
<tr><td colspan="2">(a) 100 nm L&S</td><td colspan="2">(b) 200 nm L&S</td></tr>
</table>

Figure 2. Scanning electron microscopy (SEM) images of the first and second replica molds for the transferred patterns of the (**a**) 100 nm line-and-space (L&S) and (**b**) 200 nm L&S transferred patterns.

For the 100 nm L&S pattern, the first replica mold had a line width, space width, and a height of 112, 107, and 193 nm, respectively. The second replica mold had a line width, space width, and height of 100, 104, and 158 nm, respectively. Whereas in the 200 nm L&S pattern, the first replica mold had a line width, space width, and height of 220, 190, and 236 nm, respectively. The second replica mold had a line width, space width, and height of 210, 197, and 220 nm, respectively. The variation in height from the first replica mold to the second was caused by shrinkage, which occurred during curing [20].

The durability of the second replica mold was then evaluated by repeating the transfer using the machine (Mitsui Electric Co., Ltd., Chiba, Japan) in Figure 3).

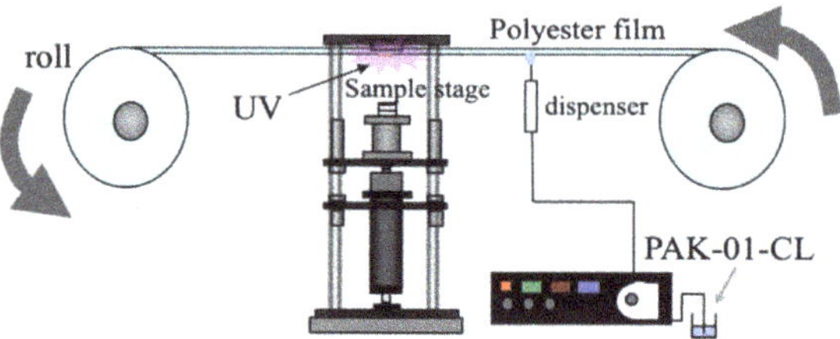

Figure 3. Transfer endurance and barge endurance device.

Repeated transcription was performed as illustrated in Figure 4. PAK-01-CL was used as the resin during the durability test. After pressurizing at 0.12 MPa and filling the resin, UV irradiation with a dose of 400 mJ/cm^2 was applied. In this case, a UV LED (ZUV -C20H (OMRON Co. Ltd., Kyoto, Japan), wavelength: 365 nm) was used for UV curing. The study was also conducted at a temperature of 21 °C.

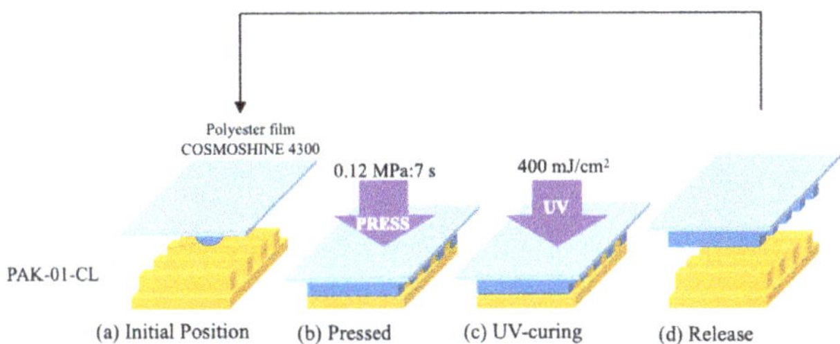

Figure 4. The process of repeated transcription.

To measure the release ability of the second replica mold, the water contact angle was measured (Drop master-701, Kyowa Interface Science Co., Ltd., Niiza City, Japan) at various imprint numbers. The water droplet volume at the time of contact angle measurement was 2 μL.

The contact angle was measured every 20 times from the 1st to 100th cycles, every 50 times from the 100th to 400th cycles, every 100 times from the 400th to 1000th cycles, every 200 times from the 1000th to 3000th cycles, and every 250 times thereafter. Moreover, the contact angle was measured five times at the center of the mold after a predetermined number of transfers. It was confirmed that the standard deviation of these five data was less than 3.0°. The structure being evaluated for durability was line-patterned; therefore, a capillary force acted while measuring the contact angle. Water was deposited and the contact angles in the x and y directions were measured after 3 s, as shown in Figure 1. Furthermore, because water flowed in the direction of the fine lines due to the capillary force, the contact angle varied depending on the measurement direction. In order to evaluate the anisotropy, contact angle measurements were made from two directions, as illustrated in Figure 5.

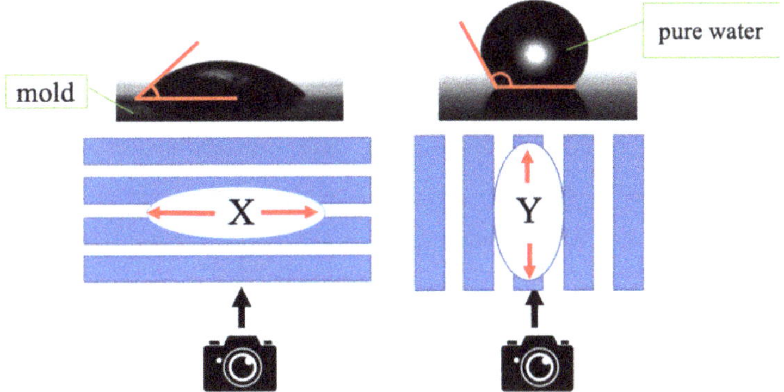

Figure 5. Measurement of the contact angle between x and y.

4. Results and Discussion

Figure 6 shows the results of the durability evaluation of the second replica mold, specifically the contact angles for the 100 nm and 200 nm L&S patterns. There were some similarities between the results obtained for these L&S widths. The contact angle was approximately 140° before the transfers, which eventually decreased with repeated transfers. Moreover, as the contact angle decreased, water began to flow along the line pattern; specifically, the contact angle in the x direction decreased to approximately 20° before it stabilized. On the other hand, the contact angle in the y direction decreased in a more linear manner. In the 100 nm L&S pattern, the contact angle in the y direction was linear after 1800 transfers with a gradient of −0.0316. In the 200 nm L&S pattern, the contact angle in the y direction was linear after 2400 transfers with a gradient of −0.0154. This gradient halved when the scale between the lines of the L&S pattern was doubled. A linear contact angle in the y direction corresponded to a smaller L&S scale, and the defect was likely to form relatively sooner.

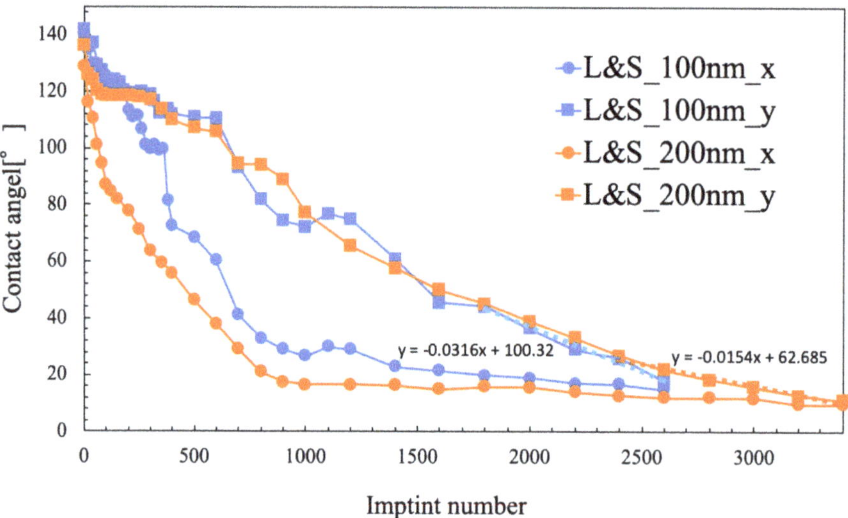

Figure 6. Contact angles after different numbers of imprints.

The experimental results indicated that the contact angle in the x direction became constant at approximately 20°. The point where the above-mentioned inclination and approximately 20° straight line intersected was predicted to be at 2600 imprints for the 100 nm L&S pattern and at 3400 imprints for the 200 nm L&S pattern. For the 100 nm L&S pattern, a defect occurred in the second replica mold at approximately transfer number 2600. For the 200 nm L&S pattern, a defect occurred in the second replica mold at approximately transfer number 3400. We categorized the lifetime of the replica mold before observing the macroscopic defect. In the line pattern mold, capillary force is generated along the line pattern, and this force spreads the water along the line direction (x direction). As a result, the contact angle observed in the x direction was lower than that measured across the line direction (y direction). Here, we assumed the mold surface energy to be constant and that the contact angle of a regular pattern (such as dots or holes) and the contact angles of x and y directions are equal. In contrast, in the case of line patterns, the direction of capillary force enhances the water spreading and exhibits lower contact angles. Furthermore, we think that the change in contact angles caused by the capillary flow indicates the future contact angle in the other direction (in this case the y direction). As shown in Figure 6, the contact angle in the x direction was saturated at 20°. This means that the contact angle in the y direction can decrease to this value (20°) without forming any macroscopic defects, because the x direction exhibited no macroscopic defects at the contact angle of 20°.

We observed the macroscopic defect as shown in Figure 7. Figure 8 shows SEM images of the transferred film surface using PAK-01-CL.

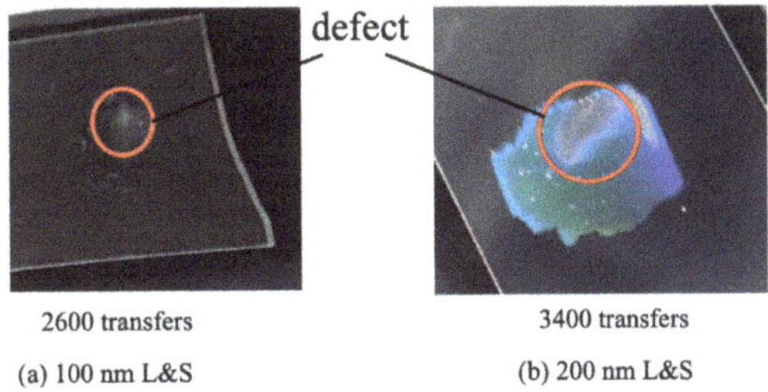

Figure 7. Images of a defect of the (**a**) 100 nm L&S and (**b**) 200 nm L&S transferred patterns.

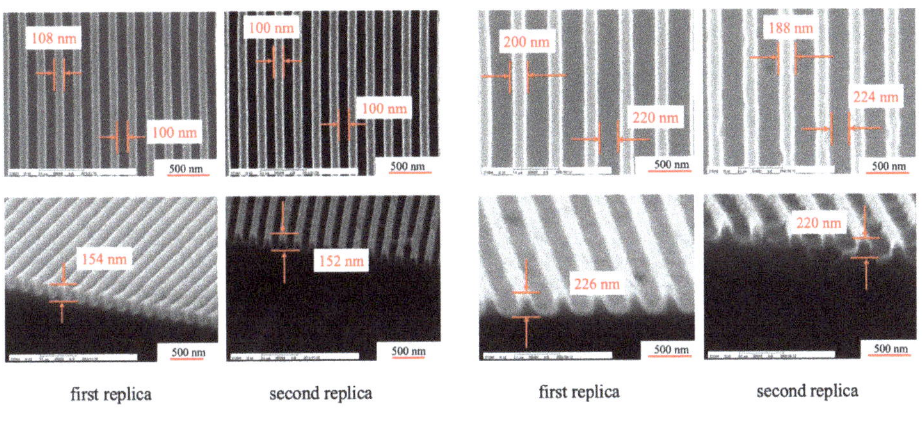

(a) 100 nm L&S

(b) 200 nm L&S

Figure 8. SEM images of the (**a**) 100 nm L&S and (**b**) 200 nm L&S transferred patterns.

In the 100 nm L&S pattern, the first transfer had a line width, space width, and a height of 108, 102, and 154 nm, respectively. The corresponding values observed for transfer number 2400 were 100, 100, 152 nm, respectively. In the 200 nm L&S pattern, the first transfer had a line width, space width, and height of 200, 220, and 226 nm, respectively. The corresponding values observed for transfer number 3200 were 188, 224, and 220 nm, respectively.

Figure 9 depicts how the contact angle decreased linearly in the y direction and approached a constant value in the x direction. The second replica mold included fluorinated materials. We considered that these release materials are gradually removed during repetition of the UV-NIL process. This is the reason for the linear decrease in the y direction (see Figure 6). Moreover, the contact angle in the x direction becomes constant at an earlier stage as compared to that in the y direction. Therefore, the intersection of x and y can be predicted. Furthermore, a visible deficit in the number of transcriptions near the intersection of x and y was observed. From these results, it is possible to predict the approximate number of transcriptions after which the defect occurs by predicting the intersection of x and y. The lifetime of the line-patterned mold could be predicted because the contact angle differed in the x and y directions. It is thought that this lifetime was the result of water that flows along the line pattern due to capillary force.

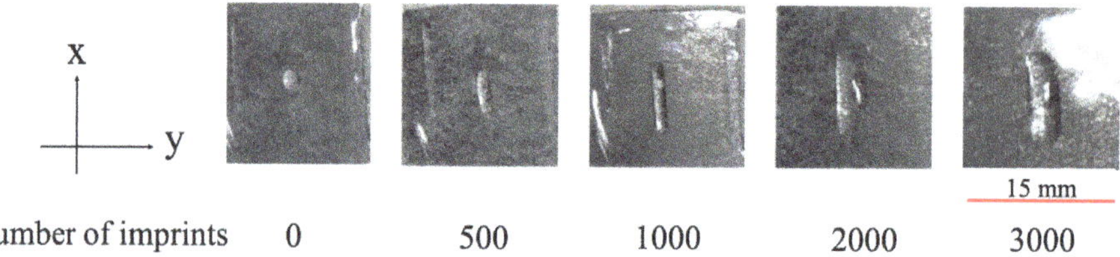

Figure 9. Water penetration distance for various imprint numbers.

Figure 10 shows the water penetration distance for various numbers of imprints. The penetration distance in the y direction increased with increasing transfer number. The penetration distance was calculated while considering the direction of the force. The values of W and H were obtained from Figure 2. μ is the viscosity of PAK-01-CL (1.002 mPa·s) and δ is the surface tension of water (7.225×10^{-5} N/mm), which is used for calculating ΔP, as expressed in Equation (5). Considering the sign of $\cos\theta$, the values provided in Table 1 were substituted into Equation (5). The penetration distance was obtained by substituting the obtained pressure ΔP into Equation (4).

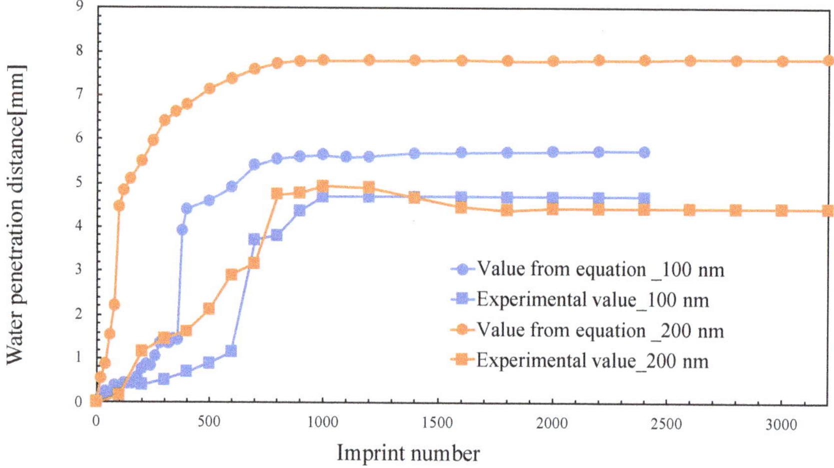

Figure 10. Water penetration distance with increasing imprint number.

Table 1. Values assigned to Equation (5) for water penetration distance.

Scale [nm]	W [nm]	H [nm]	t [s]
100	104	158	5
200	197	220	5

The measured experimental values were compared with those calculated using Equation (4). As shown in this figure, the values of water penetration distances obtained from the experiment and those evaluated from the equations were different. However, leveling off was generated after approximately 1000 imprints, which means that saturation of contact angle occurred after approximately 1000 repetitions. Therefore, the minimum requirement of lifetime prediction is approximately 1000 repetitions. Furthermore, the penetration distance between the lines was calculated. However, in practice, water on the lines exhibits the Cassie–Baxter and Wenzel states [22]. The calculated values assumed that all the water was in between the lines. In reality, the difference was caused by the presence of water above these lines as well. When the transfer was repeated, the penetration distance became constant. This was because the pressure depended on the contact angle. Moreover the theoretical and experimental values displayed similar plots; therefore, it can be said that capillary action occurs within the molds in this experiment. It can hence be concluded that the contact angle anisotropy of the molds with the line pattern exists due to this capillary action.

5. Conclusions

Ultraviolet nanoimprint lithography (UV-NIL) can be used to fabricate nanoscale patterns while ensuring a high throughput and low cost. The high-volume production of nanostructures requires the durability of the release-coated silicon mold to be excellent. Accordingly, the durability of a line-patterned mold was evaluated using a high-hardness release-agent-free resin. Capillary action occurred in the nanoscale line structure. It was observed that the contact angle decreased with repeated transfers and that the water flowed along the lines. Measuring the contact angle in the x and y directions allowed the prediction of the lifetime of the mold. This prediction method is limited to line-and-space patterns. However, using line-and-space patterns, every release coating or release material can be evaluated. In addition, the minimum number of UV-NIL imprints required to estimate the stamp lifetime is until saturation of the contact angle in the x direction occurs. We obtained the contact angles in both x and y directions vs. the number of imprints. Lifetime can be estimated from the crossing point of the x direction (saturated low contact angle value) curve and the corresponding y direction curve. This is a facile and labor-saving way. This lifetime prediction could potentially lead to a shortening of the durability evaluation time and prevention of mold breakage. If this comparison of x and y in L&S patterns, or life prediction, can be utilized, it will be possible to determine which demolding process is best for a smaller number of transfers.

Author Contributions: Conceptualization, T.M. and J.T.; methodology, T.M. and J.T.; validation, T.M. and J.T.; investigation, T.M.; materials, S.H.; writing—original draft preparation, T.M.; writing—review and editing, J.T.; project administration, J.T. All authors have read and agreed to the published version of the manuscript.

Funding: This research received no external funding.

Acknowledgments: The authors would like to thank the Toyo Gosei Corporation for providing the UV curable resins (PAK-01-CL).

Conflicts of Interest: The authors declare no conflict of interest with this manuscript.

References

1. Chou, S.Y.; Krauss, P.R.; Renstrom, P.J. Nanoimprint lithography. *J. Vac. Sci. Technol. B* **1996**, *14*, 4129–4133. [CrossRef]
2. Bender, M.; Fuchs, A.; Plachetka, U.; Kurz, H. Status and prospects of UV-Nanoimprint technology. *Microelectron. Eng.* **2006**, *83*, 827–830. [CrossRef]
3. Gilles, S.; Meier, M.; Prömpers, M.; van derHart, A.; Kügeler, C.; Offenhäusser, A.; Mayer, D. UV nanoimprint lithography with rigid polymer molds. *Microelectron. Eng.* **2009**, *86*, 661–664. [CrossRef]
4. Voisin, P.; Zelsmann, M.; Gourgon, C.; Boussey, J. High-resolution fused silica mold fabrication for UV-nanoimprint. *Microelectron. Eng.* **2007**, *84*, 916–920. [CrossRef]
5. Foncy, J.; Cau, J.-C.; Bartual-Murgui, C.; François, J.-M.; Trévisiol, E.; Séverac, C. Comparison of polyurethane and epoxy resist master mold for nanoscale soft lithography. *Microelectron. Eng.* **2013**, *110*, 183–187. [CrossRef]
6. Yamashita, D.; Taniguchi, J.; Suzuki, H. Lifetime evaluation of release agent for ultraviolet nanoimprint lithography. *Microelectron. Eng.* **2012**, *97*, 109–112. [CrossRef]
7. Jeong, G.H.; Park, J.K.; Lee, K.K.; Jang, J.H.; Lee, C.H.; Kang, H.B.; Yang, C.W.; Suh, S.J. Fabrication of low-cost mold and nanoimprint lithography using polystyrene nanosphere. *Microelectron. Eng.* **2010**, *87*, 51–55. [CrossRef]
8. Jang, E.-J.; Park, Y.-B.; Lee, H.-J.; Choi, D.-G.; Jeong, J.-H.; Lee, E.-S.; Hyun, S. Effect of surface treatments on interfacial adhesion energy between UV-curable resist and glass wafer. *Int. J. Adhesion Adhesives* **2009**, *29*, 662–669. [CrossRef]

9. Sun, H.; Liu, J.; Gu, P.; Chen, D. Anti-sticking treatment for a nanoimprint stamp. *App. Surf. Sci.* **2008**, *254*, 2955–2959. [CrossRef]

10. Mühlberger, M.; Bergmaira, I.; Klukowska, A.; Kolander, A.; Leichtfrieda, H.; Platzgummer, E.; Loeschner, H.; Ebm, C.; Grützner, G.; Schöftner, R. UV-NIL with working stamps made from Ormostamp. *Microelectron. Eng.* **2009**, *86*, 691–693. [CrossRef]

11. Tucher, N.; Höhn, O.; Hauser, H.; Müller, C.; Bläsi, B. Characterizing the degradation of PDMS stamps in nanoimprint lithography. *Microelectron. Eng.* **2017**, *180*, 40–44. [CrossRef]

12. Schleunitz, A.; Vogler, M.; Fernandez-Cuesta, I.; Schift, H.; Gruetzner, G. Innovative and Tailor-made Resist and Working Stamp Materials for Advancing NIL-based Production Technology. *Photopolymer* **2013**, *26*, 119–124. [CrossRef]

13. Garidel, S.; Zelsmann, M.; Chaix, N.; Voisin, P.; Boussey, J.; Beaurain, A.; Pelissier, B. Improved release strategy for UV nanoimprint lithography. *J. Vac. Sci. Technol. B* **2007**, *25*, 2430–2434. [CrossRef]

14. Schmitt, H.; Zeidler, M.; Rommel, M.; Bauer, A.J.; Ryssel, H. Custom-specific UV nanoimprint templates and life-time of antisticking layers. *Microelectron. Eng.* **2008**, *85*, 897–901. [CrossRef]

15. Semprebon, C.; Mistura, G.; Orlandini, E.; Bissacco, G.; Segato, A.; Yeomans, J.M. Anisotropy of Water Droplets on Single Rectangular Posts. *Langmuir* **2009**, *25*, 5619–5625. [CrossRef]

16. Schwiebert, M.K.; Leong, W.H. Underfill Flow as Viscous Flow Between Parallel Plates Driven by Capillary Action, IEEE Trans. Compon. *Packaging Manuf. Technol. C* **1996**, *19*, 133–137.

17. Huang, W.; Liu, Q.; Li, Y. Capillary Filling Flows inside Patterned-Surface Microchannels. *Chem. Eng. Technol.* **2006**, *29*, 716–723. [CrossRef]

18. Princen, H.M. Capillary phenomena in assemblies of parallel cylinders: III. Liquid Columns between Horizontal Parallel Cylinders. *J. Coll. Int. Sci.* **1970**, *34*, 171–184. [CrossRef]

19. Tsuchiya, J.; Hiwasa, S.; Taniguchi, J. Transfer durability and fidelity of hard release-agent-free replica mold by repetition of ultraviolet nanoimprint lithography. *Microelectron. Eng.* **2018**, *193*, 98–104. [CrossRef]

20. Otsuka, Y.; Hiwasa, S.; Taniguchi, J. Development of release agent-free replica mould material for ultraviolet nanoimprinting. *Microelectron. Eng.* **2014**, *123*, 192–196193. [CrossRef]

21. Hashimoto, Y.; Mogi, K.; Yamamoto, T. Release agent-free low-cost double transfer nanoimprint lithography for moth-eye structure. *IEEE-NANO* **2016**. [CrossRef]

22. Murakami, D.; Jinnai, H.; Takahara, A. Wetting Transition from the Cassie–Baxter State to the Wenzel State on Textured Polymer Surfaces. *Langmuir* **2014**, *30*, 2061–2067. [CrossRef] [PubMed]

 *nanomaterials*

Article

Development, Processing and Applications of a UV-Curable Polymer with Surface Active Thiol Groups

Manuel Müller *, Rukan Nasri, Lars Tiemann and Irene Fernandez-Cuesta *

Center for Hybrid Nanostructures (CHyN), Institut für Nanostruktur- und Festkörperphysik (INF), Universität Hamburg, Luruper Chaussee 149, 22761 Hamburg, Germany; rukan.nasri@physnet.uni-hamburg.de (R.N.); lars.tiemann@physik.uni-hamburg.de (L.T.)
* Correspondence: mamuelle@physnet.uni-hamburg.de (M.M.); ifernand@physnet.uni-hamburg.de (I.F.-C.)

Received: 22 July 2020; Accepted: 4 September 2020; Published: 14 September 2020

Abstract: We present here a novel resist formulation with active thiol groups at the surface. The material is UV curable, and can be patterned at the micro- and nanoscale by UV nanoimprint lithography. The resist formulation development, its processing, patterning and surface characterization are presented here. In addition, a possible application, including its use to modify the electrical properties of graphene devices is shown. The cured material is highly transparent, intrinsically hydrophilic and can be made more hydrophilic following a UV-ozone or an O_2 plasma activation. We evaluated the hydrophilicity of the polymer for different polymer formulations and curing conditions. In addition, a protocol for patterning of the polymer in the micro and nanoscale by nanoimprinting is given and preliminary etching rates together with the polymer selectivity are measured. The main characteristic and unique advantage of the polymer is that it has thiol functional groups at the surface and in the bulk after curing. These groups allow for direct surface modifications with thiol-based chemistry e.g., thiol-ene reactions. We prove the presence of the thiol groups by Raman spectroscopy and perform a thiol-ene reaction to show the potential of the easy "click chemistry". This opens the way for very straightforward surface chemistry on nanoimprinted polymer samples. Furthermore, we show how the polymer improves the electrical properties of a graphene field effect transistor, allowing for optimal performance at ambient conditions.

Keywords: nanoimprint lithography; polymer; formulation development; surface chemistry; click chemistry

1. Introduction

Nanoimprinting lithography (NIL) has emerged over the last two decades as a high throughput nanofabrication method and is now a consolidated technology [1–3]. The method uses a stamp, brought into physical contact with a polymer, to mold it and transfer the pattern of the stamp into the polymer. Several different approaches can be found for nanoimprinting, being the main ones thermal NIL and UV NIL. In recent years a strong industrially-driven research can be observed, pushing NIL and its applications from purely academic purposes to innovations in data storage, point of care diagnostics, electronic and graphene devices or augmented reality [4–9]. With this shift, new materials need to be developed with properties that are compatible with mass production: fast curing and imprinting, robustness, reproducibility, and patterning on flexible substrates, both, at the micro and nano scale.

NIL was initially used as a purely lithographic method, for pattern transfer [10] as a fast and cheap alternative to electron beam lithography. In recent years, a new trend has emerged towards direct patterning of functional materials, where the imprinted structures are directly used as active elements in devices. These devices find applications in several different fields, like micro and nanofluidics [11,12], biosensing [13] and DNA analysis [14], optical, nanophotonic [15] and flexible, electrical devices [16]. For example, biocompatible and three dimensional polymer structuring for cell cultivation open a promising field for imprintable polymers as an alternative to conventional processing using photoresists, which need long process times, by offering more flexibility in the processing, higher throughput and incrementing the design freedom [17].

For using patterned surfaces in devices or sensors, the surface of the cured polymer often needs to be chemically modified or activated. The functionalization endows the surface with properties different and independent from the bulk or from other untreated areas of the polymer. However, the process of functionalization often requires an activation, or a two-step modification of the surfaces with strong (and dangerous) chemicals or physical interaction [18,19]. The process is time consuming and not always compatible with the device (e.g. compatibility of biomolecules with temperature changes, chemicals or pH changes). Functional, imprintable polymers with intrinsic and easily modifiable functional groups are very attractive alternatives.

A versatile functional group for surface modifications is the thiol group due to the well-researched "click chemistry" [20] and also its strong affinity to common materials used in micro and nanoengineering, such as gold [21]. Thiols are known to undergo a nucleophilic addition with Michael systems, such as (meth)acrylates or vinyls, at room temperature [22–24], as well as radical additions. Spatially controlled reactions by ink-jet printing or masking could be used for local chemical modification of surfaces and patterns on an imprinted structure.

In this work, we have developed a resist formulation, which has functional thiol groups at the surface and in the bulk. It is easily patternable by UV NIL, both at the micro and nanoscale. In addition, it is possible to etch it selectively and homogeneously by reactive ion etching (RIE) using an oxygen plasma. The surface properties have been characterized by measuring the contact angle for different formulations and curing conditions. The thiol groups have been observed by Raman spectroscopy, confirming their presence at the surface and in the bulk, and also after an O_2 etching. As a first application, we show the possibilities for surface click chemistry on the polymer surface. And, as a second possible application, we show how the polymer, just placed on top of a graphene device, changes its electrical properties and allows for operation at ambient conditions, without ultra-high vacuum.

2. Methods

2.1. Curing of the Material

Some of the results shown in this work were obtained by curing the polymer with a monochromatic light emitting diode (LED) lamp (365 nm) under a home-made CO_2 flow box. These include those shown in Sections 3.2, 4.2.2 and 4.3.2 (nanometric lines). The rest of the results were obtained using a dedicated UV-NIL equipment (EVG 501, EV Group, Sankt Florian am Inn, Austria), with a controlled atmosphere in the chamber, and a pulsed, broadband xenon lamp (Model LH-810 Spiral Lamp 107-mm, type C spectrum, 190 nm cutoff, power 105 mW/cm^2, XENON Corporation, Wilmington, MA, USA) for exposure.

2.2. Contact Angle

The time-dependent water contact angle (WCA) measurements presented in Figure 2 were done in a Krüss DSA25 (Drop Shape Analyzer) (Krüss GmbH, Hamburg, Germany) in the sessile drop configuration. It should be noted that the dynamic contact angle in Figure 3 differs significantly from the sessile drop configuration, resulting in comparably smaller contact angles.

2.3. Dynamic Contact Angle

The setup to measure the dynamic water contact angle of the measurements presented in Figure 1 is a self-made device with a video recording system and a manual syringe for solvent application. A drop of water is placed on the flat cured surface of one of the prototypes. A video is recorded while the water droplet volume increases by slowly pressing out the water out of the syringe. The recorded video is then analysed image by image and the WCA is calculated assuming a perfectly spherical shape. For each droplet, the WCA was calculated seven times along the growth of the droplet. On each surface three different measurements were conducted and the average over all was used to compare the different surfaces.

2.4. SEM

SEM images were obtained in a Crossbeam 550L system from Zeiss (Oberkochen, Germany). A low acceleration voltage (4 kV) was used, to avoid charging of the polymers and for an optimal image quality. Some of the images were obtained with the sample tilted at 30°.

2.5. Raman Spectroscopy

The measurements are performed in a self-built setup consisting of a Ti-U inverted microscope (Nikon, Tokio, Japan) with a dichroic mirror > 538.9 nm, a 100x objective with NA = 0.90, a DPSS green laser (532 nm, 35 mW at focal point) and an Andor spectrometer (1200 l/mm grating, iDus 420 CCD detector, Andor, Belfast, United Kingdom). A confocal pinhole (105 µm) is used to reduce the out-of-focal-plane signal.

2.6. Surface Modification for Contact Angle Measurements

The UV-ozone surface activation was done in a UV-ozone Cleaner 144AX-220, from Jelight Co Inc. (Irvine, CA, USA). The O_2 plasma treatment was done in an oven (Technics Plasma 100-E, 2.45 GhZ, Technics Plasma GmbH, Kirchheim bei München, Germany) at 180 W, 2 mbar, with O_2 flow.

2.7. Graphene Transfer and Graphene Field Effect Transistor (GFET) Sample Fabrication

Monolayer graphene synthesized by chemical vapour deposition (CVD) on copper foil was released from its growth medium, cleaned and transferred to a *p*-doped Si substrate with a 100 nm surface layer of SiO_2 for electrical insulation. A 200 µm by 20 µm Hall bar structure was patterned by standard optical lithography, O_2 plasma etching and PVD of Au for the Ohmic contacts. A detailed description of the fabrication process can be found in [25]. The device was then transferred to a ceramic chip carrier and wire-bonded using a combination of ultrasonic wedge bonding (for the chip carrier pads) and conductive silver paint (for the graphene contacts). The *p*-doped substrate was also connected and acted as the back gate. Suitably sized polymer flakes were peeled off of the fabrication substrate, carefully oriented and placed over the graphene Hall bar under a microscope using a pair of tweezers. The flakes were fabricated by curing in contact with a non-polar, fluorinated stamp in the EVG 501, with a pulsed, broadband UV xenon lamp with 105 mW/cm^2 for 19 s with 2000 mJ/cm^2 total dose.

2.8. Electrical Measurements

We employed a standard lock-in method for the electrical measurements. The chip carrier with the sample was connected to a probe station and remained exposed to atmosphere/ambient air at room temperature during the measurements. The internal oscillator of a SR830 lock-in amplifier (Stanford Research Systems, Sunnyvale, CA, USA) was connected to a 100 Mega-Ohm resistor in series to the sample to generate a constant low frequency (ac) current of 2 nA (13.33 Hz) in the Hall bar. Changes in the conductivity were measured with the lock-in's differential amplifier between opposite contact pairs where the graphene was either covered by the polymer or exposed to ambient conditions. These contact pairs also acted as source and drain contacts (two-terminal measurements). The gate voltage, applied with a source measurement unit (2400 SMU, Keithley Cleveland, OH, USA), was swept in steps of 100 mV with an integration time of 1 s.

3. Development of the Material Formulation

The goal of this work was to develop a resist formulation capable of fulfilling industrial needs for rapid prototyping of micro and nanostructures for with hydrophilic surface properties, while incorporating chemical functionality into the final device. Balancing the ratio of different functional groups in the formulation, a material, which is further capable of post-cure functionalization via thiol-ene click chemistry on the surface, was obtained. Patterning of the functional polymer is straight-forward by direct curing by UV light, as sketched in Figure 1.

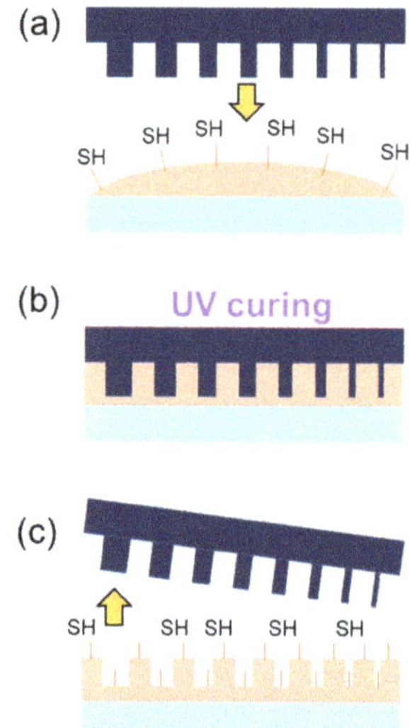

Figure 1. Micro and nanostructuring of the thiol polymer. The polymer (orange) with thiol-groups (-SH) intrinsic to its formulation is drop casted or spin coated on the substrate (blue) (**a**). By bringing in contact with a stamp (dark blue), the polymer fills the structure and is cured with UV light (**b**). After curing, the stamp is released and the polymer surface has the positive structure of the stamp and the functional thiol group (**c**).

3.1. Base Formulation

When developing a new UV-curable resist, different material characteristics need to be addressed and taken into account, such as nature of the cross linking, curing speed, hardness, flexibility, oxygen insensitivity during curing, hydrophobicity and many more. By choosing the right monomers, many of those aspects can be addressed simultaneously or need to be balanced. For instance, while a fast curing speed is often required for fast processing, the generally used radical curing mechanism results in an oxygen sensitivity, so the presence of oxygen will prevent the cross-linking reaction to occur. On the other hand, processing requirements for the imprinting step, such as low forces during the separation step may contradict final application needs such as hydrophilicity for micro- and nanofluidic behavior. Our goal was to develop a material formulation that shows both, a reduced oxygen sensitivity and an increased hydrophilicity as well as the possibility of post-functionalization by having active surface groups.

To accomplish this, in collaboration with our colleagues at Micro Resist Technology GmbH (Berlin, Germany), we used a variety of formulations, as summarized in Table 1. We started off with commercial monomers. We used an aliphatic polyurethane acrylate (PUA) as the base resin (formulation 1), since it is widely used within the community and easy to work with. In addition, it gives overall good hardness and adhesion towards polymeric surfaces. In order to introduce the hydrophilicity, two different trifunctional monomers were used, as detailed in Table 1. Each one introduced flexibility and hygroscopicity to the overall formulation that needed to be balanced with the amount of PUA used. In particular, reducing the oxygen sensitivity and introducing free thiol groups was accomplished by using hydrophilic trifunctional thiol monomers. In order to reduce the inevitable dark reaction occurring in thiol-ene based formulations, proprietary stabilizers were used. With those commercially available monomers, first the hydrophilicity was adjusted (formulations 2–5) and later on, the oxygen sensitivity using the thiol-ene analog (formulation 6).

Table 1. Summary of the used monomers for prototyping a hydrophilic UV-curable formulation with post-functional properties.

		Formulation Name					
		1	2	3	4	5	6
Employed Formulation Components	**Main Impact on Cured Material**	**Monomer Concentration Range (%)**					
Aliphatic polyurethane acrylate [a] (PUA)	Hydrophobicity, High cohesion, hardness, and glass transition temperature	100	90	70	50	0	25
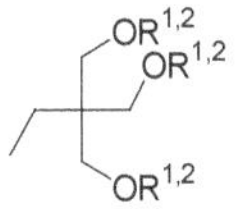 R^1 = 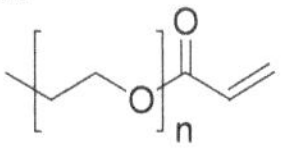	High flexibility, high hydrophilicity	0	10	30	50	100	30
R^2 = 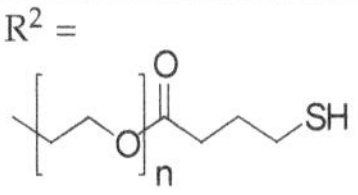	Flexibility, oxygen insensitivity, post-functionalization of thiol groups	0	0	0	0	0	40
Stabilizers							>1
Photoinitiator		2–5	2–5	2–5	2–5	2–5	2–5

[a] Structure not defined by material supplier.

Formulation 6 showed optimal material properties, including the targeted presence of free thiol groups on the surface. The formulation is easily patternable by UV-NIL. The cured material is transparent, and the surface hydrophilic. The presence of the thiol groups was confirmed by Raman spectroscopy, and their functionalization by click chemistry as shown later in this article.

3.2. *Hydrophilicity and Hygroscopicity of Base Formulation*

The hydrophilicity of the polymer surface is determined by the concentration and chemistry of the surface groups. Since water is a polar liquid, surfaces with higher concentrations of polar groups will have lower contact angles than those with non-polar groups. Thus, to make the polymer more or less hydrophilic, we have tuned the ratio between the hydrophobic polyurethane acrylate (PUA) and the trifunctional hydrophilic acrylate (see Table 1, formulations 1–5). In addition, the molecular length of these monomers can be changed for fine-tuning, by increasing the amount of ethylene oxide (EO) moieties per monomer in the synthesis.

We investigated the hydrophilicity of several formulations. For this, we measured the advancing dynamic water contact angle (WCA) on the resulting materials after UV curing under different conditions, as shown in Figure 2. In particular, we varied the chemical composition of the formulations: the weight percentage of the hydrophilic trifunctional monomer and the length of the polyglycol groups, indicated by the relative ethylene oxide (EO) amount in the synthesis of the monomer.

Figure 2, red circles, shows the WCA for the formulations shown in Table 1. To our surprise, there was very little change in the hydrophilicity of the different surfaces when curing the material under inert, non-polar conditions. We used a CO_2 atmosphere, but N_2 or Argon should work similarly. Neither changing the monomer, nor the ratio changed the resulting hydrophilicity significantly. WCAs of around 70–75° were observed, even though up to 50% of hydrophilic monomer was used, as indicated by the red squares in the graph in Figure 2. Only when using 100% of the trifunctional hydrophilic monomer, a WCA of 60° was reached (formulation 5).

On the other hand, when curing under hydrophilic (polar) conditions (using a non-structured OrmoComp® stamp, plasma activated; Micro Resist Technology, Berlin, Germany) the overall WCA dropped to roughly 54°, almost indifferent of the hydrophilic content of the formulation. The red triangles in the graph in Figure 2 show the WCA of formulations 2, 3 and 4, cured under a polar surface. Varying the length of the ethylene glycol chain while maintaining the monomer concentration at 50%, the WCA dropped down to 47°. This can be seen also in the graph, where the black, red, green and blue markers correspond to different variations of formulation 4, with increasing amount of EO moieties per monomer (i.e., with different lengths of the polyglycol groups).

One possible explanation for this strong dependence on the curing conditions could be related to the free energy diffusion of the monomers before and during curing. In a non-polar surrounding, the hydrophobic (i.e., non-polar) parts of the formulation will arrange at the surface and will be fixated there upon irradiation. In the second case, the behavior is the opposite, and the polar parts will move to the surface when the uncured resist is in contact with the polar stamp. This could also explain why there is small variation in the contact angle when comparing different monomer compositions.

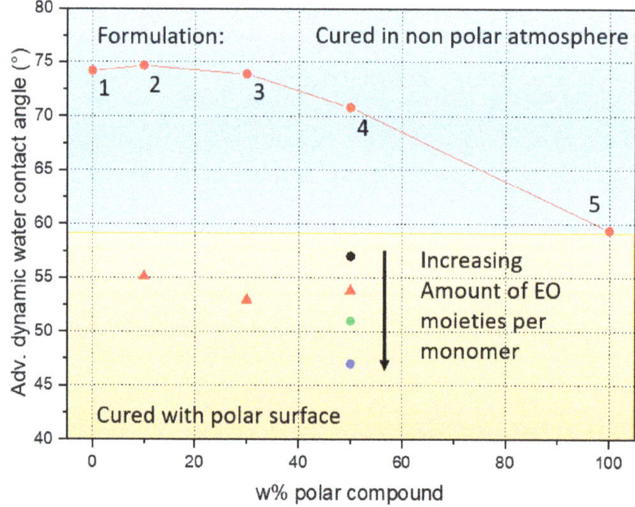

Figure 2. Evaluation of the effect of the chemical composition of the thiol polymer prototypes on its hydrophilicity, measured by the advancing dynamic contact angle. Formulations with a varying polar components content are shown in red dots for a 10%, 30% and 50%, all with WCA below 57°, when cured in contact with a polar surface (i.e., using a non-structured Ormocomp stamp).

The hydrophilicity does not come without the drawback of a certain amount of water absorption into the polymer (hygroscopicity). Since the water absorption might lead to swelling, the hygroscopicity should be accounted for if the feature size is critical. Formulations 1–5 (see Table 1) have been cured and the bulk layers (roughly 500 µm thick) were tested for water absorption. Depending on the amount of hydrophilic monomer, the water absorption varied between 1 to 30 w% depending on the length of the ethylene glycol moieties and the amount of polar monomer added to the formulation.

4. Thiol-Based Polymer

The results from those tests were used for defining the optimal material (formulation 6), which balances both the hydrophilicity and hygroscopicity, offering a good compromise: the cured material has a water uptake of less than 2%, while still having enough excess thiol groups on the surface for post functionalization. This material has a dynamic contact angle of 40°, which we have measured on samples cured without and with a stamp, functionalized with a monolayer of fluorosilanes.

4.1. Hydrophilicity and Hygroscopicity of Thiol Formulation

We have observed a non-steady behavior of water sessile drops on the surface of the cured film. Due to the interaction of the water with the surface a decrease in the observed contact angle can be seen. As the cured polymer shows a certain degree of hygroscopicity, it absorbs water and the polymer/water interface changes over time. The time dependent change of the contact angle of a water drop on untreated, pristine thiol polymer is shown in Figure 3 (black line). This pristine sample, fabricated by spin coating and curing with 2000 mJ/cm^2 in an EVG 501 shows an initial contact angle of 47°, which decreases in two minutes by 7°.

A surface activation using UV-Ozone or oxygen plasma oxidizes the organic groups at the surface, leaving more hydroxyl (–OH) groups exposed, increasing the polymer's hydrophilicity. Figure 3 (red line) shows the change of the WCA for a polymer layer treated in a UV-Ozone plasma for four minutes. The treatment reduces the contact angle by 10° with respect to the untreated, pristine sample. Its time development is very similar to the pristine sample, indicating that the UV-Ozone results in only a chemical surface modification.

A similar measurement, obtained after treating a sample in oxygen plasma (Technics Plasma 100-E) for four minutes is shown in Figure 3 (blue line). The initial wetting resulted in a higher contact angle than that of the pristine sample, but dropped within ten seconds below the value obtained for the UV-Ozone treated sample. The WCA becomes constant after a couple of minutes with a value of 23°.

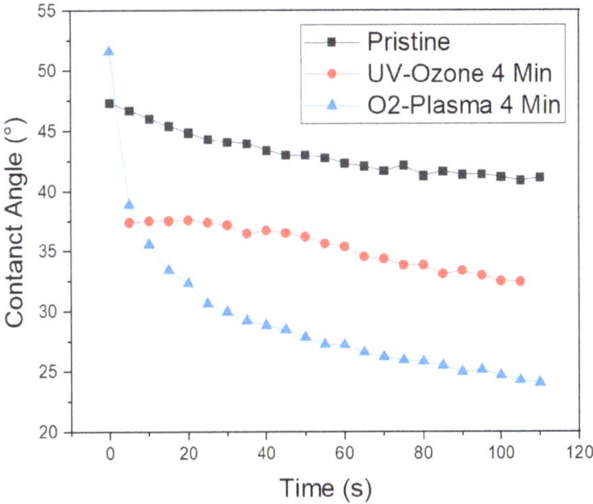

Figure 3. Contact angle measured on an untreated, pristine cured polymer layer (black line), a sample after a four-minute UV-Ozone treatment (red line) and a sample after a four-minute O$_2$ plasma activation (blue line). The WCA was measured every five seconds, and the graph shows its evolution along time for the three different samples.

The increased hydrophilicity after UV-ozone and O$_2$ plasma are stable for few hours—we have measured no change in the WCA two hours after the treatment. But the effect of the treatment is not permanent. We have observed that, after few days, the measured contact angle for the UV-ozone treated samples increases and the initial value is recovered after one week, when storing the samples at room temperature and ambient conditions. The regain of the surface properties can be attributed to a rearrangement of functional groups at the surface. This entropy-driven process will result in the diffusion and rearrangement of the short polymeric chains at the surface: the polar groups will be oriented towards the bulk and the more hydrophobic groups to the surface, realizing an equilibrium matching the atmosphere [26]. The O_2 plasma changes more drastically the contact angle, and leads to a more durable change, since it also introduces a nanoscopic surface roughness. And, in hydrophilic surfaces, higher roughness leads to lower WCA [27,28].

Since the O_2 plasma etches the polymer (in our case, <5 nm/min), it should be used with precaution, especially when dealing with features with small dimensions, or when the roughness of the surface is a critical parameter.

4.2. Presence of Thiol Groups

4.2.1. Raman Spectroscopy

When curing (meth)acrylates in the presence of thiols, two competitive reactions occur: the homopolymerization of the acrylates and the photo-initiated thiol-ene reaction. Since both reactions have different mechanisms (chain growth vs. step growth [22,23]) which take place on the same time-scale, an homogenous distribution of thiol groups in the bulk and on the surface after polymerization can be expected. In order to confirm the presence of thiol groups on the surface of the cured polymer, and also inside the bulk material, we performed Raman spectroscopy on the cured films to check for the vibrational peak corresponding to the thiol group (R-SH) at 2575 cm^{-1}.

Figure 4 compares the Raman scattering spectrum of different polymers. The red line in the graph corresponds to a thiol-containing polymer (Table 1, formulation 6), and the black line to an exemplary organic, thiol-free polymer with similar formulation. The R–SH vibrational peak is apparent only in the thiol polymer measurements, confirming the presence of thiol groups just in the thiol polymer prototype.

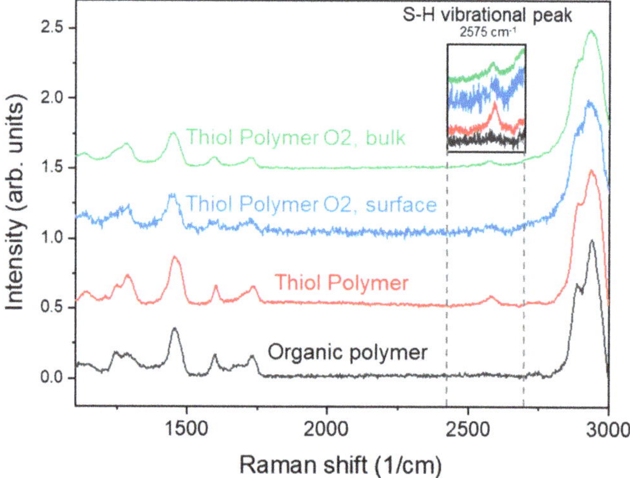

Figure 4. Raman scattering of the polymer samples. The graph shows the signal obtained on a non-thiol polymer ("Biopolymer", an organic polymer, used as control) (black line). The red line shows the signal from a thiol polymer film. The blue and green signals correspond to a film etched using an O_2 plasma, and measured at the surface and the bulk respectively. The inset shows a detail of the peak at 2575 cm^{-1}, which corresponds to the R–SH vibrational mode of the thiol groups.

We also selectively etched the thiol-polymer with an O_2 Plasma (Technics Plasma 100-E) for four minutes and measured the Raman scattering spectrum again (Figure 4, blue line). The etching process and surface modification after the oxygen plasma treatment does not lead to a significant difference in the signal. Furthermore, by changing the focal point of the laser, we can obtain the signal from the bulk material. Figure 4 (green line) shows the signal obtained from the bulk material, where the 2575 cm^{-1} peak is still present, confirming that there are thiol groups also in the bulk material. As mentioned before, the thiol groups are part of the cured, cross-linked monomers, and thus, are an intrinsic part of the bulk polymer. After polymer etching, the new groups from the newly created surface will now become superficial.

4.2.2. Surface Functionalization

As it has been described previously in the literature [20], excess thiol groups can be used to functionalize surfaces. This allows one, for example, to alter their wetting behavior. Or, more interestingly, to make them active and selective to certain chemical groups, for example to covalently attach anchor groups that selectively bind to specific biomolecules. As a proof of concept, we observed the change in the surface properties after performing a Michael addition on the cured thiol polymer, as sketched in Figure 5a–c. For this, a mixture of a fluorinated acrylate ($C_{12}H_5F_{15}O_2$) and trimethylamine (1 mmol solution) was incubated on the cured polymer film for 24 h at room temperature (Figure 5b). The solution was then washed with isopropanol and dried (Figure 5c). During the incubation, the free thiol groups reacted with the acrylates, as sketched in (c), which resulted in an increase in the hydrophobicity. Since this reaction is not radical driven, but a nucleophilic attack, no side reactions or grafting of the polymer chain are expected. We have seen that the water contact angle of the polymer surface after curing was 40°, as shown in Figure 5d, and that it increased up to 77° after the functionalization, as shown in (e).

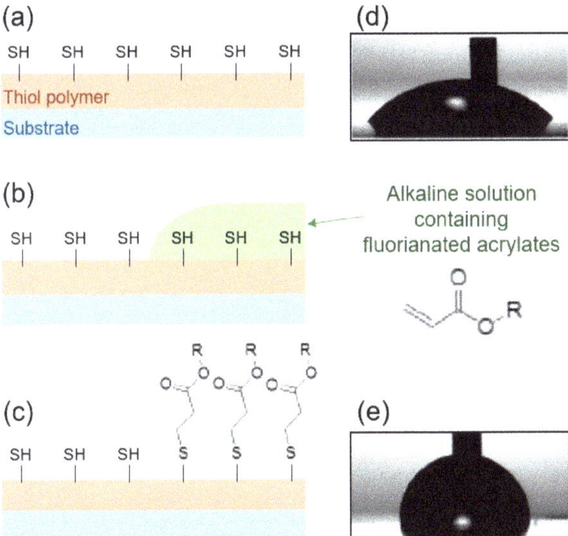

Figure 5. Surface chemistry on the polymer using the thiol functional groups. An imprinted thiol polymer sample (**a**), which has surface-active thiol groups, is incubated with a solution of fluorinated acrylate and trimethylamine (**b**) for 24 h. The molecules selectively react with the thiol functional groups in a Michael addition (**c**). The contact angle of the untreated sample is 40° (**d**), while after the chemical surface modification, it increases to 77° (**e**).

Full coverage with a self-assembled monolayer of fluorosilanes typically results in WCA of >100° [29]. Thus, since we observe a WCA of 77°, we assume that we don't have a full coverage. We estimated it in the range of 40–60%, which is probably reflecting the amount of thiol groups at the surface. To increase the coverage, a larger amount of surface thiol groups should be present at the surface; for this, a different formulation should be used, at the expense of different material properties.

This experiment proofs that the thiol groups at the surface are active and reactive to chemical modifications, opening the way to direct, simple surface modifications using wet chemistry, since the process can be performed with other "clickable" monomers.

4.3. Material Properties and Patterning

4.3.1. General Properties

The cured polymer is transparent to visible light and has a slight yellowish color, resulting from the used photoinitiator, when cured in thick layers of several micrometers. The transmittance is 99.5% for visible light ($\lambda \geq 380$ nm), measured on a 1.6 µm thick layer of cured polymer on a glass slide. The transmission spectrum for the visible range can be seen in Figure S1 in the supplementary material. The polymer, owing to its acrylate and thiol-ene chemistry, shows typical behavior against solvents and acids. It is resistant to isopropanol, but not to acetone. It shows a high optical transparency (>99%) in the visible range with a refractive index of 1.486. Its intrinsic hydrophilicity (water contact angle 40°) is low enough for spontaneous microfluidic behavior in open channel microfluidics and top surface functional groups (R–SH, and R–OH) have been reported. These and other general properties of the investigated prototype liquid and cured resist are given in Table 2.

Table 2. General properties of the polymer before and after curing.

Uncured Material	
Viscosity	330 m Pas
Mold compatibility	Silicon, Glass, Ormostamp
Curing atmosphere	Oxygen free
Curing dose	>2000 mJ/cm^2
Cured material	
Refractive index	1.486
Transparency ($\lambda \geq 350$ nm)	>95%
Transparency ($\lambda \geq 380$ nm)	>99%
Surface functional groups	SH, OH
Contact Angle	40°
Layer thickness (spin coating, 3000 rpm)	5.0 µm
Layer thickness (spin coating, 6000 rpm)	1.9 µm
Layer thickness (drop casting)	20–70 µm

4.3.2. Patterning

The patterning of the UV-resist is sketched in Figure 1a–c: the material is drop casted or spin coated on the substrate (a), then, a hard stamp is brought into contact, and the polymer is cured under UV light (b); the stamp is separated from the substrate, and the pattern is replicated on the polymer (c). The typical thickness of the drop-casted imprinted layers is around 70 µm. Drop casting is a manual process, so the droplet volume and changes in the viscosity (associated to temperature fluctuations) cause a variation in the final layer thicknesses. Reproducible and controlled layer thicknesses are achieved by spin coating. Film thickness of 1.9 and 5.0 µm were obtained for spin speeds of 3000 and 6000 rpm, respectively.

The developed material prototype can be processed both by spin coating or drop casting. Different materials, like silicon, glass or polymers (e.g. PMMA and polycarbonate, PC) are suitable substrates. The adhesion to the substrate can be improved by cleaning and activation by UV-Ozone or oxygen plasma; an adhesion promoter is recommended for inorganic substrates. Since it is solvent free, no pre-baking step is necessary, simplifying the processing and improving the throughput and compatibility to different substrates. The polymer is cross-linked by exposing it to monochromatic or broadband UV light. A minimum curing dose of 2000 mJ/cm^2 is recommended. This leads to a high transparency and hardness.

Gas-impermeable stamps should be used (like glass or silicon), since the curing chemistry is sensitive to oxygen. Thus, PDMS stamps should be avoided. All imprints shown in this work have been performed with hard, non-gas permeable stamps, made of silicon or cured Ormostamp®, to avoid any interfering oxygen at the surface or diffusion into the polymerizing liquid. Further high intensity UV irradiation was used to overcome the effect of possible residual oxygen. The low viscosity of the uncured material enables imprinting at room temperature with low pressure: the weight of the stamp or the sample is usually sufficient to start filling the cavities, which is then followed by capillary forces that result in a complete filling with the liquid resist within seconds. In order to decrease the releasing forces during the separation of the stamp and the cured resist, the stamps were coated with an anti-sticking layer of fluorosilanes by chemical vapor deposition [29].

In this work, we have used an EVG Imprint tool (EVG 501 customized for UV and thermal imprinting) for sample imprinting. The tool allows for chamber evacuation before the NIL process, to remove oxygen from the liquid polymer. The chamber is evacuated to 0.1 mbar for 20 s to degas the polymer and then purged again with air. The exposure is done with a pulsed, broadband UV Xenon lamp (190 nm cutoff) with 105 mW/cm^2 for 19 s, resulting in a dose of 2000 mJ/cm^2. Other tests with a non-pulsed, low energy density, monochromatic UV LED lamp (365 nm) and a custom-made flow box to locally create an inert gas atmosphere worked as well.

Figure 6 shows scanning electron microscope (SEM) images of thiol-polymer imprinted structures with lateral dimensions in the micro- and nanoscale. Dots with diameters ranging from 1.2 to 11 µm can be seen in (a) and meandering lines, 2 µm wide and 4.8 µm deep in (b). Nanostructures and lines with gradually smaller widths from 200 to 75 nm are shown in (c). (d) shows tilted SEM images (30°) of a combination of a 20 µm wide line and several straight and meander structures, 5 µm wide, which cross section can be seen in (e). The top view of these identical lines can be seen in (f). The smallest feature sizes obtained were 75 nm wide and 200 nm height with a pitch of 150 nm. It is noteworthy that the shown resolution of the nanostructures is mainly limited by the stamp; imprinting of smaller features should be possible. The developed thiol-polymer prototype allows for simple, easy and reproducible combination of structures with large and small dimensions without problems.

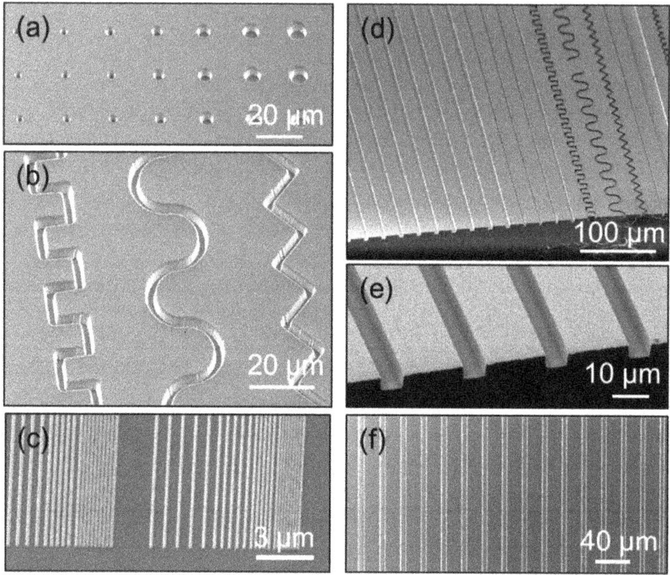

Figure 6. SEM images of micro- and nano-imprinted structures are shown. The structures vary from dots, lines, meander structures, and rely only on the quality of the structures in the stamp. Micro structures with different shapes and configurations are shown in (**a,b,d–f**) (dots, meandering lines, straight lines). And also lines with nanometric dimensions, down to 75 nm wide, as shown in (**c**).

4.3.3. Etching

The thiol polymer prototype can be structured and removed by reactive ion etching (RIE). Standard pattering techniques leave a residual layer after imprinting. In order to remove this layer for etching and pattern transfer, RIE has been proven to be a suitable method. We have used an ICP-RIE system (SI 500 RIE, from Sentech) to test the etching properties for standard pure O_2 and SF_6-based recipes, typically used for etching micro and nano structures into polymers or silicon, respectively. The results for the etching rate for the polymer for the two different processes are shown in Table 3. The etching rate of 0.4 µm/min obtained for the O_2 plasma is comparable to that for other typical organic polymers etched under similar conditions. The resulting surface is very smooth, as can be seen in Figure S2a in the supplementary information.

Table 3. Etching of the thiol polymer by RIE. Two typical RIE recipes used to remove or structure polymers (O_2 etch) and silicon (Si etch) were evaluated.

Process	ICP Power	HF Power	Gas	Pressure (Pa)	Flow (sccm)	Temp. (°C)	Etch Rate (µm/min)
O_2 etch	150 W	50 W	O_2	0.5	30	21	0.4
Si etch (anisotropic)	400 W	15 W	O_2, SF_6, C_4F_8	1.0	5, 50, 70	3	0.2 (Si 0.6)

Polymers are often used as masks for etching or are part of more complex devices which undergo further processing. For this, we have tested the polymer also against recipes commonly used to etch silicon. We used a typical recipe, where SF_6, C_4F_8 and O_2 are mixed (details shown in Table 3). With this process, the polymer is etched at 0.2 µm/min and silicon at 0.6 µm/min. This gives a selectivity of 1:3. This recipe leads to rougher surface even for short etching times (100 s), as can be seen in Figure S2b in the supplementary information. We also tested the polymer in an O_2 SF_6 RIE (without the passivant, C_4F_8), but obtained worse selectivity against silicon, and rougher surfaces (see Figure S2c,d).

It should be noted that in our reactors chamber, which contains aluminum, we observe some times Al sputtering on the surface during some of the RIE processes—especially those with high chamber pressures; small clusters of aluminium are deposited on the polymer surface and act as a mask for the etching, resulting in grass-like polymer structures (see Figure S2d in the supplementary material). This "grass effect" has been reported in the literature and is not unique to this polymer [30].

4.4. Application: Electrical Measurements on Graphene

One further application to exploit the intrinsic properties of the newly developed material lays in the field of graphene field effect transistors (GFET). Graphene is a two dimensional, highly conductive material. Since its conductivity is strongly dependent on the purity of the material and on the surface properties, even minor amounts of adhered molecules will result in doping and thereby a strong change in the properties of the GFET. One critical adhering molecule can be water from the humidity in the air. In order to obtain reliable data under ambient conditions, water needs to be removed. Thus, electrical measurements of GFETS are typically performed at ultra-high vacuum conditions. We have found that placing a cured layer of the polymer developed here (formulation 6) directly on top of the GFET results in a decreased doping, as the resist will absorb the moisture due to its hygroscopicity.

We compared the conductivity of a graphene field effect transistor patterned on a silicon oxide substrate with and without the cured thiol polymer adhered on top, as shown in the sketch in Figure 7a and in the image of an actual device in Figure 7b, using a back gate to control the charge carrier type and concentration. This is a standard device configuration to quantify doping in graphene [31]. By measuring the resistance as a function of the gate voltage, as shown in the graph in Figure 7c, the maximum resistivity indicates the voltage needed to empty the conductance band and completely fill the valence band of the graphene; this value is the so-called charge neutrality point (CNP). Intrinsically, the CNP of graphene lies at zero volts and shifts due to doping (for example due to adhered molecules or humidity on the surface). Figure 7c, green line, shows the resistance for a sweep in the gate voltage for the uncovered graphene transistor at ambient conditions (room temperature, and a humidity of 40–50%). The curve can be better seen in the inset. The red curves in the graph show the signal of the same GFET, patterned using the same graphene flake, where it was locally covered by the polymer (as sketched and shown in the drawing and image in Figure 7a,b). The presence of the polymer leads to a very strong shift of the CNP from higher positive gate voltages towards the zero bias region. We confirmed this effect in two additional graphene devices.

Since the measurements were done at ambient conditions in clean GFETs, just water is expected to be adhered to the surface of the graphene flake. Thus, the adsorbed water molecules dope the graphene (p-doping), shifting the CNP towards higher gate voltages [32]. As a control experiment, we covered a different GFET with a non-hygroscopic polymer (Ormostamp®). These measurements on "Device 2" are shown in the bottom panel of Figure 7c. The resistance as a function of the gate voltage measured for the uncovered GFET Device 2 "at air"is shown in the green line and shows a similar behavior as Device 1. The base resistance differs since the intrinsic doping can vary between different devices. As we put a layer of cured Ormostamp® on the graphene, a change in the signal is observed, as can be seen in the orange line. However, the hygroscopic effect of the thiol polymer on graphene as seen for Device 1 is not reproduced with this other polymer.

These results underline the relevance of the hygroscopic properties of the thiol polymer and its possible application for example for graphene based electronics, which could function at ambient conditions [33]. In addition, this also enables the operation close to the CNP. Changes in the charge carrier density around the CNP result in the strongest changes in conductivity, enhancing the graphene's sensitivity for small perturbations, which is highly desirable in detectors.

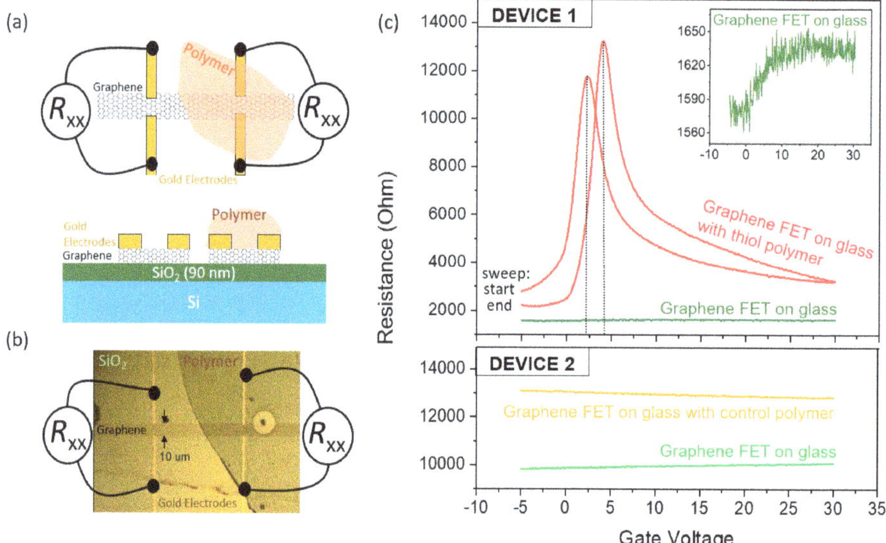

Figure 7. (**a**) Graphene field effect transistor (GFET). A graphene flake is structured on a SiO$_2$/Si substrate, and gold electrodes patterned on top, to act as electrical contacts for measurements. Part of the GFET is locally covered with a cured piece of thiol-polymer, as sketched in the top and side views. The p-doped silicon substrate acts as a back gate, which is electrically isolated from the graphene Hall bar by a layer of SiO$_2$ (bottom panel). The upper part of the Hall bar is covered by polymer whereas the lower part remains exposed. (**b**) Photo of an actual device, where the area on the right side is darker due to polymer deposited on top. (**c**) Measurement of the two-terminal resistances of the GFET device as a function of a gate voltage. The upper graph (Device 1) shows the results for one of the measured devices. The green line corresponds to the graphene "at-air". The red lines correspond to the measurement on the graphene covered with the thiol-polymer, for a sweep from −5 V to 30 V, and from 30 V to −5 V, as marked in the graph. The lower graph (Device 2) shows the results of a similar experiment performed in another different graphene FET device, where instead of the thiol polymer, a layer of a non-hygroscopic polymer (Ormostamp) is used to cover the graphene, as control experiment.

5. Conclusions

A polymer prototype containing thiol functional groups and an intrinsic hydrophilicity was presented. The polymer is UV curable, and can be patterned at the micro and nanoscale by UV nanoimprint lithography. The thiol groups present at the surface can be directly used for chemical functionalization.

The presented formulation uses PUA as base resin, and two different trifunctional monomers. To adjust the hydrophilicity of the product, the contact angle of cured material made of different mixtures was measured after curing the material. It was observed that the percentage in weight of the hydrophilic components played a minor role, and that mainly the curing conditions determined the final contact angle.

A mixture of 25% PUA, 35% trifunctional monomer and 40% trifunctional thiol monomer was proposed as the optimal formulation. This material has a contact angle of 40° and a small, intrinsic hygroscopicity. The polymer is highly transparent in the visible range (T > 99%) and has a refractive index of n = 1.486. Structures with lateral dimensions spanning from several microns down to 75 nm were imprinted by UV-NIL. The dry etching (RIE) of the polymer was investigated for O$_2$ and O$_2$ + C$_4$F$_8$ + SF$_6$ plasmas, and the selectivity to silicon was shown to be 1:3.

The presence of the thiol groups in the optimal formulation was confirmed by Raman spectroscopy. The peak corresponding to the vibrational mode corresponding to the thiol group at 2575 cm^{-1} was present in the cured polymer samples, on the surface and also in the bulk material, confirming that the thiol groups are an intrinsic part of the polymer. Even after O_2 plasma etching, the Raman peak corresponding the thiol groups is present, confirming the compatibility of the polymer with standard O_2 plasma processing.

To further prove the presence of thiol groups at the surface, and to show one example of the potential of the polymer for applications requiring surface chemistry (e.g., "click chemistry"), a Michael addition was performed. A solution of fluorinated acrylate in trimethylamine was incubated on a flat, cured polymer surface, resulting in a significant increase in the water contact angle as a result of the change in the surface chemistry, showing a surface coverage in the range of 40 to 60%.

As another example of application, we have shown that the cured thiol polymer has a strong effect on the electrical properties of graphene. We measured the CNP of a GFET device at ambient conditions and showed a large shift towards zero when the graphene was coated with a thin layer of polymer, as compared to a non-coated one. Since the polymer is hygroscopic, it absorbs the residual layer of water (originating from ambient humidity) from the graphene surface, leading to electrical properties of the GFET similar to those which can only be obtained under vacuum conditions when the devices are not coated with the polymer layer.

The work presented in this paper shows the potential and versatility of the proposed polymer, which is easy to process and pattern at the micro and nanoscale, and which allows for a variety of applications yet to be explored. Our preliminary results pave the way for using the polymer for click chemistry processes and opens an application window for simple imprinting of complex nanostructures and consecutive easy chemical surface modifications.

Supplementary Materials: The following are available online at http://www.mdpi.com/2079-4991/10/9/1829/s1, Figure S1: Polymer transparency and Figure S2: Surface roughness.

Author Contributions: Conceptualization, methodology, I.F.-C. and M.M.; Validation and investigation M.M., L.T., R.N.; writing—original draft preparation M.M. and I.F.C.; writing—review and editing M.M., I.F.-C., L.T., R.N.; funding acquisition, I.F.-C. All authors have read and agreed to the published version of the manuscript.

Funding: This project has received funding from the European Research Council (ERC) under the European Union's Horizon 2020 research and innovation program (grant agreement No 714073). This research was also partially funded from the European Union's Horizon 2020 research and innovation program under grant agreement No 646260.

Acknowledgments: Authors would like to thank the team at micro resist technology GmbH in general, and Mirko Lohse and Manuel Thesen in particular, for the help with the prototype development and the support with the measurements.

Conflicts of Interest: The authors declare no conflict of interest.

References

1. Chou, S.Y.; Krauss, P.R.; Zhang, W.; Guo, L.; Zhuang, L. Sub-10 nm imprint lithography and applications. *J. Vac. Sci. Technol. B Microelectron. Nanometer Struct. Process. Meas. Phenom.* **1997**, *15*, 2897–2904. [CrossRef]

2. Chou, S.Y.; Krauss, P.R.; Renstrom, P.J. Imprint lithography with 25-nanometer resolution. *Science* **1996**, *272*, 85–87. [CrossRef]

3. Schift, H. Nanoimprint lithography: An old story in modern times? A review. *J. Vac. Sci. Technol. B Microelectron. Nanometer Struct. Process. Meas. Phenom.* **2008**, *26*, 458–480. [CrossRef]

4. Krauss, P.R.; Chou, S.Y. Nano-compact disks with 400 Gbit/in2 storage density fabricated using nanoimprint lithography and read with proximal probe. *Appl. Phys. Lett.* **1997**, *71*, 3174–3176. [CrossRef]

5. Giannone, D.; Dortu, F.; Bernier, D.; Johnson, N.P.; Sharp, G.J.; Hou, L.; Khokhar, A.Z.; Fürjes, P.; Kurunczi, S.; Petrik, P.; et al. NIL fabrication of a polymer-based photonic sensor device in P3SENS project. In Proceedings of the SPIE Photonics Europe, Brussels, Belgium, 16–19 April 2012; SPIE: Brussels, Belgium, 2012; Volume 8435.

6. Schowengerdt, B.T. Virtual and Augmented Reality Systems and Methods. U.S. Patents US9791700B2, 17 October 2017.

7. Mackenzie, D.M.A.; Smistrup, K.; Whelan, P.R.; Luo, B.; Shivayogimath, A.; Nielsen, T.; Petersen, D.H.; Messina, S.A.; Bøggild, P. Batch fabrication of nanopatterned graphene devices via nanoimprint lithography. *Appl. Phys. Lett.* **2017**, *111*, 193103. [CrossRef]

8. Müller, A.; Vu, X.T.; Pachauri, V.; Francis, L.A.; Flandre, D.; Ingebrandt, S. Wafer-Scale Nanoimprint Lithography Process Towards Complementary Silicon Nanowire Field-Effect Transistors for Biosensor Applications. *Phys. Status Solidi (A)* **2018**, *215*, 1800234.

9. Kam, A.P.; Seekamp, J.; Solovyev, V.; Cedeno, C.C.; Goldschmidt, A.; Torres, C.S. Nanoimprinted organic field-effect transistors: Fabrication, transfer mechanism and solvent effects on device characteristics. *Microelectron. Eng.* **2004**, *73–74*, 809–813. [CrossRef]

10. Messerschmidt, M.; Greer, A.; Schlachter, F.; Barnett, J.; Thesen, M.W.; Gadegaard, N.; Grutzner, G.; Schleunitz, A. New Organic Photo-Curable Nanoimprint Resist ≪mr-NIL210≫ for High Volume Fabrication Applying Soft PDMS-Based Stamps. *J. Photopolym. Sci. Technol.* **2017**, *30*, 605–611. [CrossRef]

11. Fernandez-Cuesta, I.; Laura Palmarelli, A.; Liang, X.; Zhang, J.; Dhuey, S.; Olynick, D.; Cabrini, S. Fabrication of fluidic devices with 30 nm nanochannels by direct imprinting. *J. Vac. Sci. Technol. B* **2011**, *29*, 06F801. [CrossRef]

12. Fernandez-Cuesta, I.; West, M.M.; Montinaro, E.; Schwartzberg, A.; Cabrini, S. A nanochannel through a plasmonic antenna gap: An integrated device for single particle counting. *Lab Chip* **2019**, *19*, 2394–2403. [CrossRef]

13. Merino, S.; Retolaza, A.; Trabadelo, V.; Cruz, A.; Heredia, P.; Alduncin, J.A.; Mecerreyes, D.; Fernández-Cuesta, I.; Borrisé, X.; Pérez-Murano, F. Protein patterning on the micro- and nanoscale by thermal nanoimprint lithography on a new functionalized copolymer. *J. Vac. Sci. Technol. B Microelectron. Nanometer Struct. Process. Meas. Phenom.* **2009**, *27*, 2439–2443. [CrossRef]

14. Esmek, F.M.; Bayat, P.; Perez-Willard, F.; Volkenandt, T.; Blick, R.H.; Fernandez-Cuesta, I. Sculpturing wafer-scale nanofluidic devices for DNA single molecule analysis. *Nanoscale* **2019**, *11*, 13620–13631. [CrossRef]

15. Guo, L.J. Recent progress in nanoimprint technology and its applications. *J. Phys. D: Appl. Phys.* **2004**, *37*, R123–R141. [CrossRef]

16. Shao, J.; Chen, X.; Li, X.; Tian, H.; Wang, C.; Lu, B. Nanoimprint lithography for the manufacturing of flexible electronics. *Sci. China Ser. E: Technol. Sci.* **2019**, *62*, 175–198. [CrossRef]

17. Koitmäe, A.; Müller, M.; Bausch, C.S.; Harberts, J.; Hansen, W.; Loers, G.; Blick, R.H. Designer Neural Networks with Embedded Semiconductor Microtube Arrays. *Langmuir* **2018**, *34*, 1528–1534. [CrossRef] [PubMed]

18. Fukai, R.; Dakwa, P.H.R.; Chen, W. Strategies toward biocompatible artificial implants: Grafting of functionalized poly(ethylene glycol)s to poly(ethylene terephthalate) surfaces. *J. Polym. Sci. Part A: Polym. Chem.* **2004**, *42*, 5389–5400. [CrossRef]

19. Sugiyama, K.; Kato, K.; Kido, M.; Shiraishi, K.; Ohga, K.; Okada, K.; Matsuo, O. Grafting of vinyl monomers on the surface of a poly (ethylene terephthalate) film using Ar plasma-post polymerization technique to increase biocompatibility. *Macromol. Chem. Phys.* **1998**, *199*, 1201–1208. [CrossRef]

20. Claudino, M.; Jonsson, M.; Johansson, M. Utilizing thiol–ene coupling kinetics in the design of renewable thermoset resins based on d -limonene and polyfunctional thiols. *RSC Adv.* **2014**, *4*, 10317–10329. [CrossRef]

21. Xue, Y.; Li, X.; Li, H.; Zhang, W. Quantifying thiol–gold interactions towards the efficient strength control. *Nat. Commun.* **2014**, *5*, 4348. [CrossRef]

22. Natarajan, L.V.; Shepherd, C.K.; Brandelik, D.M.; Sutherland, R.L.; Chandra, S.; Tondiglia, V.P.; Tomlin, D.; Bunning, T.J. Switchable Holographic Polymer-Dispersed Liquid Crystal Reflection Gratings Based on Thiol–Ene Photopolymerization. *Chem. Mater.* **2003**, *15*, 2477–2484. [CrossRef]

23. Rydholm, A.E.; Bowman, C.N.; Anseth, K.S. Degradable thiol-acrylate photopolymers: Polymerization and degradation behavior of an in situ forming biomaterial. *Biomaterials* **2005**, *26*, 4495–4506. [CrossRef]

24. Lutolf, M.P.; Hubbell, J.A. Synthesis and Physicochemical Characterization of End-Linked Poly(ethylene glycol)-co-peptide Hydrogels Formed by Michael-Type Addition. *Biomacromolecules* **2003**, *4*, 713–722. [CrossRef]

25. Lyon, T.J.; Sichau, J.; Dorn, A.; Zurutuza, A.; Pesquera, A.; Centeno, A.; Blick, R.H. Upscaling high-quality CVD graphene devices to 100 micron-scale and beyond. *Appl. Phys. Lett.* **2017**, *110*, 113502. [CrossRef]

26. Teare, D.; Ton-That, C.; Bradley, R. Surface characterization and ageing of ultraviolet-ozone-treated polymers using atomic force microscopy and x-ray photoelectron spectroscopy. *Surf. Interface Anal. Int. J. Devoted Dev. Appl. Tech. Anal. Surf. Interfaces Thin Films* **2000**, *29*, 276–283. [CrossRef]

27. Daryaei, E.; Tabar, M.R.R.; Moshfegh, A. Surface roughness analysis of hydrophilic SiO_2/TiO_2/glass nano bilayers by the level crossing approach. *Phys. A Stat. Mech. Appl.* **2013**, *392*, 2175–2181. [CrossRef]

28. Wenzel, R.N. Resistance of solid surfaces to wetting by water. *Ind. Eng. Chem.* **1936**, *28*, 988–994. [CrossRef]

29. Schift, H.; Saxer, S.; Park, S.; Padeste, C.; Pieles, U.; Gobrecht, J. Controlled co-evaporation of silanes for nanoimprint stamps. *Nanotechnology* **2005**, *16*, S171–S175. [CrossRef]

30. Gogolides, E.; Constantoudis, V.; Kokkoris, G.; Kontziampasis, D.; Tsougeni, K.; Boulousis, G.; Vlachopoulou, M.; Tserepi, A. Controlling roughness: From etching to nanotexturing and plasma-directed organization on organic and inorganic materials. *J. Phys. D: Appl. Phys.* **2011**, *44*, 174021. [CrossRef]

31. Lee, H.; Paeng, K.; Kim, I.S. A review of doping modulation in graphene. *Synth. Met.* **2018**, *244*, 36–47. [CrossRef]

32. Yang, Y.; Brenner, K.; Murali, R. The influence of atmosphere on electrical transport in graphene. *Carbon* **2012**, *50*, 1727–1733. [CrossRef]

33. Lafkioti, M.; Krauss, B.; Lohmann, T.; Zschieschang, U.; Klauk, H.; Klitzing, K.V.; Smet, J.H. Graphene on a Hydrophobic Substrate: Doping Reduction and Hysteresis Suppression under Ambient Conditions. *Nano Lett.* **2010**, *10*, 1149–1153. [CrossRef] [PubMed]

 *nanomaterials*

Article

Novel Concept of Micro Patterned Micro Titer Plates Fabricated via UV-NIL for Automated Neuronal Cell Assay Read-Out

Mirko Lohse [1,*], Manuel W. Thesen [1], Anja Haase [2], Martin Smolka [2], Nerea Briz Iceta [3], Ana Ayerdi Izquierdo [3], Isbaal Ramos [4], Clarisa Salado [4] and Arne Schleunitz [1]

[1] Micro Resist Technology GmbH, Köpenicker Str. 325, 12555 Berlin, Germany;
    m.thesen@microresist.de (M.W.T.); a.schleunitz@microresist.de (A.S.)
[2] Joanneum Research Materials, Institute for Surface Technologies and Photonics, 8160 Weiz, Austria;
    anja.haase@joanneum.at (A.H.); Martin.Smolka@joanneum.at (M.S.)
[3] TECNALIA, Basque Research and Technology Alliance (BRTA), Mikeletegi Pasealekua 2,
    20009 Donostia-San Sebastián, Spain; Nerea.briz@tecnalia.com (N.B.I.); ana.ayerdi@tecnalia.com (A.A.I.)
[4] Innoprot, Parque Tecnológico de Bizkaia, Edificio 502, Primera Planta, 48160 Derio-Bizkaia, Spain;
    iramos@innoprot.com (I.R.); csalado@innoprot.com (C.S.)
*   Correspondence: m.lohse@microresist.de

Abstract: The UV-nanoimprint lithography(UV-NIL) fabrication of a novel network of micron-sized channels, forming an open channel microfluidic system is described. Details about the complete manufacturing process, from mastering to fabrication in small batches and in high throughput with up to 1200 micro titer plates per hour is presented. Deep insight into the evaluation of a suitable UV-curable material, mr-UVCur26SF is given, presenting cytotoxic evaluation, cell compatibility tests and finally a neuronal assay. The results indicate how the given pattern, in combination with the resist, paves the way to faster, cheaper, and more reliable drug screening.

Keywords: nanoimprint lithography; R2R UV-NIL; neuronal cell assay

Citation: Lohse, M.; Thesen, M.W.; Haase, A.; Smolka, M.; Iceta, N.B.; Ayerdi Izquierdo, A.; Ramos, I.; Salado, C.; Schleunitz, A. Novel Concept of Micro Patterned Micro Titer Plates Fabricated via UV-NIL for Automated Neuronal Cell Assay Read-Out. *Nanomaterials* **2021**, *11*, 902. https://doi.org/10.3390/nano11040902

Academic Editor: Konstantins Jefimovs

Received: 8 February 2021
Accepted: 29 March 2021
Published: 1 April 2021

Publisher's Note: MDPI stays neutral with regard to jurisdictional claims in published maps and institutional affiliations.

1. Introduction

Cell assays are an essential part of biochemical research, pharmaceutics and diagnostics. The interaction between cells and towards external stimulus, such as a potential drug, can give insight into the potential toxicity or pharmaceutic effect of such chemicals [1–5]. Some of the cell-based assays to determine the toxic potential of drugs are reactive oxygen species measurements (ROS) [6], Caspase activation [7], life cell imaging, [8] DNA damage [9], and neurite outgrowth [10]. Traditionally, such cell assays have been performed on flat surfaces using commercially available microtiter plates (MTP) fabricated with thermoplastic materials like polystyrene (PS), polymethyl methacrylate (PMMA) or cyclo-olefin copolymers (COC). However, it has been shown that the surface topography can play a crucial role in the tendency of cells to adhere to the surface [11], grow in defined directions or shapes [12], and even, in terms of stem cells, in which way cells differentiate [13–15]. While most of these patterns range from a few microns to several tens of microns [16–20], several examples show that even nanosized features [21–23] can influence the cell behavior. One particular class of cells of great interest is neurons. Their use in drug screening has been reported extensively and different commercial devices are available [24]. The toxic effects of the drugs on neurons could be measured by analyzing the neuron projection characteristics (neurites), such as neurite average length, neurite root number and neurite branching. On flat state-of-the-art microtiter plates the entanglement and overlapping of the different neurites make automated read out difficult and unreliable. However, there is a multitude of different ways for improving here. While some use micro or nano patterned surfaces to help cells to orientate them along the pattern [25,26], one common way is to block neuron cell cores with a physical barrier, which only the neurites can pass [27–30]. Thereby, the reaction of single neurites to external stimulus can be observed. We believe that with a well-chosen patterned surface, the deviations appearing during the analyses of flat surface neuron clusters can be reduced and automated read-out can be improved. Looking at the literature, most examples of patterned surfaces rely on a thermal process, either by injection molding or thermal nanoimprint lithography (NIL) [31], limiting the choice of material and the functionality beyond the one given by the pattern. UV-curable materials and the use of UV-NIL offer the chance to broaden the range of materials that can be used and incorporate intrinsic functionalities such as hydrophilicity of the material or functional groups on the cured surface. Further, UV-NIL is a highly scalable process. It was shown that by UV-NIL both nano- and micro-featured patterns can be replicated and even whole lab-on-a-chip devices [32].

In our approach, we want to overcome the limitations of standard cell assays on flat surfaces, by proposing to employ micropatterning processes for an innovative micro titer plate design (a) driving the growth of neurites to targeted 2-dimensional channels and thereby avoid crowding and entanglement of neurons and protecting them from removal during washing; and (b) limiting the number of neurons in a given spot, making image analysis more feasible and reliable (see Scheme 1). Further, the connection of each well by the 2-dimensional channels creates an open channel microfluidic system that enables selective functionalization of the channel walls at the base area of each well and channel. Our aim for this work is to show that UV-NIL is a highly feasible method to fabricate such patterns in high volume by roll-to-roll (R2R) fabrication. Therefore, we have chosen a UV-curable resist that is known for its compatibility with continuous R2R fabrication and show its compatibility with the demands of cell testing [33].

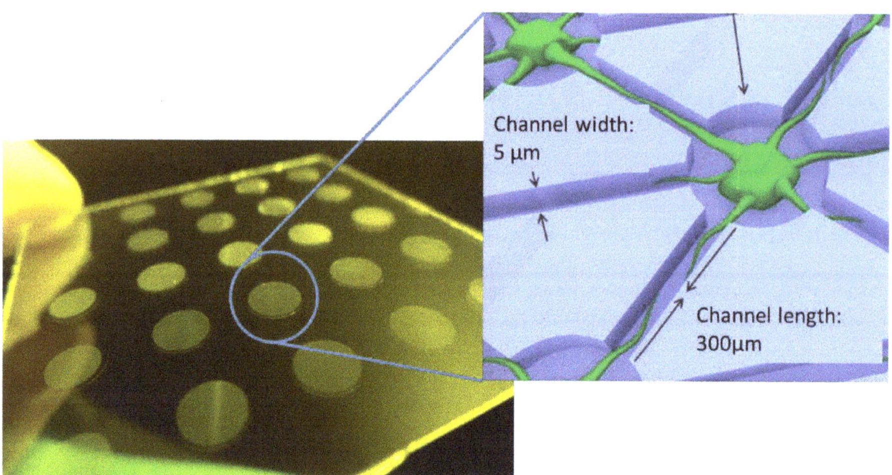

Scheme 1. Micro patterned bottom foil of a micro titer plate and schematic design of the underlying concept of cell separation.

2. Materials and Methods

2.1. Mastering

Silicon wafers from Siegert Wafer GmbH (Aachen, Germany) were used and oxygen plasma treated (900 W, 1 min) before coating. Negative type photoresist SU-8 3000 was received by Kayaku Advanced Materials Inc. (Westborough, MA, USA) and soft mold UV-PDMS KER-4690 was bought from Shin-Etsu Chemical Co., Ltd. (Tokio, Japan). The mr-Dev 600 and OrmoStamp was offered by micro resist technology GmbH (Berlin, Germany). The 1H,1H,2H,2H-Perfluorooctyl-trichlorosilane (CAS number [78560-45-9], F13-OTCS) was received from ABCR GmbH (Karlsruhe, Germany) and used without further purification.

Plasma activation of the silicon wafer was done on a PS300 Semi Auto oxygen plasma oven from PVA Tepla AG (Wettenberg, Germany). Polymer masters were fabricated using a MA6 mask aligner from Süss MicroTec SE (Garching, Germany). Optical micrographs of imprinted structures were recorded using a BX51M microscope from Olympus K.K. (Tokio, Japan) equipped with a Color View Soft Imaging System. SEM images were performed on an Inspect F50 from FEI Company (Hillsboro, OR, USA).

The fabrication of thick SU-8 based negative type photoresist polymer masters (see Scheme 2) has been published in detail before [34]. Briefly, a 17 μm thick SU-8 3000 layer was patterned with the corresponding mask on a MA6 mask aligner and developed using mr-Dev 600. Afterwards, an antisticking layer (ASL) was applied via chemical vapour deposition (CVD) of F13-OTCS following a standard procedure derived from literature [35]. Two different orientations (A and B in Scheme 2) of lithography masks were used depending on the final use, resulting in either recessed or free-standing microfluidic channels. Exact horizontal side wall accuracy could be achieved with the established photolithography process (see Figure 1). It is noteworthy that a full 96 well plate could not be mastered as this exceeded the dimension of a 6-inch wafer, which was the limit of the applied mask aligner. Nevertheless, a master with approximately 60 micro titer wells was fabricated.

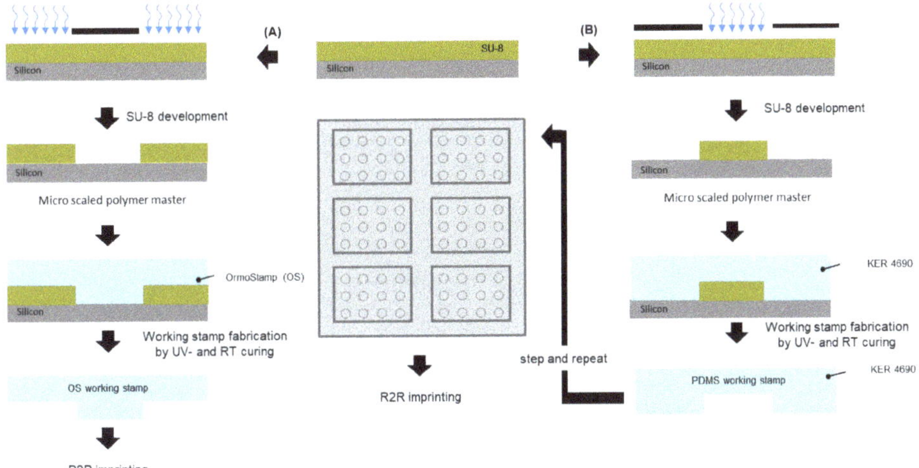

Scheme 2. Schematic representation of the fabrication process to obtain the micro patterned bottom foil either by plate-to-plate (P2P) (**A**) or roll-to-roll (R2R) (**B**).

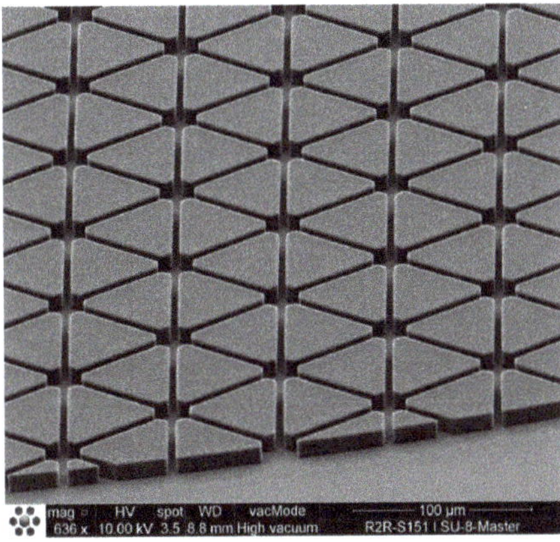

Figure 1. SEM image of the SU-8 polymer master showing the microfluidic network. At the bottom, parts of the feeding channel are visible.

Orientation A with free trenches was directly transferred into a hard-working stamp typically applied in plate-to-plate imprint processes with extremely high requirements on pattern fidelity. Therefore, in a reverse-NIL process a defined amount of OrmoStamp [36] was poured on the ASL equipped polymer master and covered with a glass backplate. After the liquid resin had spread and filled the cavities of the master, the photo-polymerization process was initialized by UV-exposure in the MA6 mask aligner ($D = 2000$ mJ/cm^2). Afterwards, a post exposure bake (150 °C, 30 min) and coverage with a CVD F13-OTCS ASL (similar to above) gave the final OrmoStamp working stamp used as such for the plate-to-plate (P2P) imprinting. Working stamps with sizes up to 7×7 cm^2 (25 micro titer wells) have been fabricated and used for batch imprinting. With larger working stamp sizes the forces needed to separate the rigid polymer master from the rigid glass like OrmoStamp were too high and resulted in destruction of the polymer master.

Orientation B, with free standing trenches, was used for the fabrication of a soft stamp intermediate for the production of large-scale shims for the R2R manufacturing. Therefore, a defined amount of silicone prepolymer, enough to cover the full wafer, was poured onto the polymer master with orientation B. Once the cavities of the SU-8 were completely filled with the KER-4690, which was monitored via microscope imaging, the UV-PDMS was exposed in the MA6 mask aligner ($D = 2000$ mJ/cm^2) to activate the crosslinking. The detachment of the UV-PDMS stamp copy from the SU-8 master was conducted 12 h after exposure and storage at room temperature. The process of replicating SU-8 polymer masters with PDMS is well known and has already been applied for the manufacturing of other microwell applications using generic PDMS [37–39]. With this approach a full 6″ wafer was replicated and used for the step and repeat fabrication for the R2R shims.

2.2. P2P UV-NIL Imprints

Foil substrates PET (Melinex, 125 μm) used for P2P and R2R test imprints were purchased from Pütz Folien. Foils were precleaned with isopropanol and, if needed, activated by a short oxygen plasma (200W, 45 s). UV-NIL resist mr-UVCur26SF was offered by micro resist technology GmbH (Berlin, Germany). mr-UVCur26SF is a solvent-free and fast curing acrylate formulation designed for R2R and Inkjet applications [33]. Its high pattern stability, high double bond conversion (>95%), low shrinkage (about 8%) and low auto fluorescence made it an ideal candidate for the given application.

Two different ways for processing the P2P imprint have been performed. The precleaned and preactivated foil was either spin coated in a manual tabletop spin coater at 3000 rpm for 60 s to achieve a thin layer of resist or a film was prepared using a 10 μm thick doctor blade. For imprinting, the film was brought into contact with a working stamp bearing the negative of the desired pattern. By manually laying the stamp from one side of the imprint area to the other, a slow filling of the cavities was achieved. In terms of the microscopic pattern aimed at in the publication, the filling of the pattern could be observed through an optical microscope. When complete filling was achieved, curing of the UV-curable resist was either conducted using a tabletop device (CNI v.2.0 supplied by NIL Technology ApS, Lyngby, Denmark) or using an MA6 from Süss MicroTec (Garching, Germany). Optical micrographs of the imprinted structures were recorded using a BX51M microscope from Olympus K.K. (Tokio, Japan) equipped with a Color View Soft Imaging System.

2.3. Converting to R2R UV-NIL

Continuous roll-to-roll (R2R) UV-NIL has been proven to be an efficient way to fabricate microfluidic devices [32]. The shim fabrication for large area imprinting derived from the previously described PDMS replica of the SU-8 polymer master (orientation B, Scheme 2). Those 6″ working stamps were used in a proprietary step and repeat process by Inmold ApS (Horsholm, Denmark) [40], followed by the sputtering of a layer of chromium and aluminum, 3–4 nm and 50 nm, respectively, at the premises of Joanneum Research. Finally, the shim was coated with an ASL to minimize the releasing energies during high throughput fabrication [41]. The obtained finished shim was mounted on a magnetic roller inside the R2R machine with a minimal seam between the two ends. The R2R imprint experiments were realized on a customized Basecoater R2R-UV-NIL imprinting machine from Coatema Coating Machinery GmbH (Dormhagen, Germany). The resist was applied via a double column gravure coating resulting in a comparable thick film of resist (approximately 20 μm, not optimized) on a PET-foil and the pattern was imprinted using the metalized roller shim. Web speeds between 0.1 m/min and 2 m/min were tested and curing was conducted using an LED-source (395 nm).

Final imprints were investigated for air bubble defects by standard optical microscope.

2.4. Cytotoxicity Validation

In this paper, we have used several cell lines provided by INNOPROT (Derio, Spain); kidney tubular epithelial cells (KTEC, Cat N: P10647), conjunctive epithelial cells (Epit. conj., Cat N: P10870), Dermal endothelial cells (Skin. Endo., Cat N: P10861), hepatic endothelial cells (Hepatic. Endo., Cat N: P10652), and two cell lines obtained from ATCC (Virginia, USA); bone osteosarcoma epithelial cells (U2OS, Cat N: ATCC HTB-96) and embryonic kidney 293 cells (HEK293, Cat N: ATCC CRL-1573).

Biocoating and cell seeding was performed manually. Visualization of cells and neurons was done using Zeiss Axio Observer Fluorescence microscope equipped with Apotome 2 system (structured illumination system by Zeiss, Germany).

To investigate the compatibility of the used UV-NIL, resist cytotoxicity tests following UNE-EN-ISO 10993-5: 2009 have been conducted. Briefly, following the processing guidelines of the resist, a layer of the material was applied on PET foil (precleaned with isopropanol) via spin coating (3000 rpm, 2 acc, 30 sec) resulting in a 700 nm thick film, followed by curing under an inert (CO_2) atmosphere in a MA6 mask aligner (D = 2000 mJ/cm^2). A defined surface area of the cured resist was then placed in contact with an incubation medium (complete Eagle's minimal essential medium (EMEM)) for 24 h at 37 °C (surface area to sample volume was defined as 6 mL/cm^2).

The American Type Culture Collection CCL 171 cell line was cultured in EMEM medium supplemented with 10% fetal bovine serum (FBS) and 1% antibiotic penicillin/streptomycin (under standard cell-culture conditions (37 ± 2 °C and of 5% CO_2) until the cells reached the confluent state. For the assay 1×10^4 CCL-171 cells were seeded in complete EMEM in 96-well plates and incubated at 37 °C and 5% CO_2. The following day, cells were incubated with the sample extracts (100, 75, 50 and 25%) and control extracts for 24 h. Cell viability was quantified by means of the WST-1 test. WST-1 reagent is added to each well and 4 h later the optical density of this solution in each well is measured at a wavelength of 450 nm using a microplate reader (BioTek Instruments, Inc. (Winooski, VT, USA), Powerwave XS). Each measurement was conducted five times in parallel. Additionally, a negative control, a positive control and a blank were included in the assay. As for the ISO guidelines, high density polyethylene was used as the negative control and polyurethane film containing 0.1% zinc diethyldithiocarbamate (ZDEC) as the positive control.

According to ISO 10993-5: 2009, a sample is considered toxic if cell viability at the highest concentration analyzed (namely, 100%) is below 70% of viability. In this test the overall performance of different materials can be tested in terms of impurities that can migrate from the bulk into the incubation medium. It is noteworthy, that no material properties, such as its cell adhesion properties were investigated at this stage. Further, no material related toxicity (e.g., certain functional groups on the surface) was evaluated at this stage.

To investigate the biocompatibility of the used UV-curable material, cell assays were conducted on a cured plain film, which was treated with plasma, or used pristine, as well as coated with different types of biopolymers (collagen and poly-D-lysine, PDL). Tested cell types were kidney tubular epithelial cells (Monkey KTEC, P10647 Innoprot), conjunctive epithelial cells (human Epit conj, P10870 Innoprot), hepatic endothelial cells (human

Hepatic endo, P10652 Innoprot), skin endothelial cells (human skin endo, P10861), bone osteosarcoma epithelial cells (human U2OS, HTB-96 ATCC) and human embryonic kidney 293 cells (human HEK, CRL-1573 ATCC). Cell lines were tested on their cell adhesion as well as proliferation on the given surface at 24 h and after 72 h being in contact to the material surface (incubation temperature 37 °C). For such assay, 20,000 cells/well were cultivated on the given substrate and after a specific time (24 h and 72 h), the cell viability is determined by WST-8 colorimetric method. At the end of the desired incubation period WST-8 reagent (Cell Counting Kit-8, Dojindo) was added. Cells were then incubated at 37 °C for 60 to 90 min at 37 °C in the dark. The absorbance of the formazan product at the wavelength of 450 nm was quantified using a microplate reader. The data was expressed as a percentage of viable cells compared to the survival of a control group (untreated cells). If the given material surface treatment resulted in a cell viability of 80% or higher compared to a control (pristine PS), the cured material passed the test and was considered non-toxic for this cell line.

2.5. Neuronal Growth

For the neuronal cell assays PMMA foil (100 μm) was used as a substrate. Bottomless micro titer plates (material polystyerene) were bought from Greiner and double sided tape were purchased from BiFlow Systems GmbH.

Double sided duct tape was applied by lamination to the bottomless MTP and circular cut-outs were obtained via laser cutting using a picosecond laser (MicroStruct Vario, 3D Micromac AG, Chemnitz, Germany) with a wavelength of 355 nm. Biocoating and cell seeding was performed manually. Visualization of cells and neurons was done using Zeiss Axio Observer Fluorescence microscope equipped with Apotome 2 system (structured illumination system by Zeiss, Oberkochen, Germany).

Neuronal cell assays were performed on the patterned surface to evaluate the improvement of neurite outgrowth in the produced devices. They were fabricated in a reverse NIL process on PMMA as substrate. A defined amount of liquid resin was applied on the stamp and brought into contact with the substrate. After a complete filling was observed under the microscope, the resist was cured on a CNI 2.0 tool (365 nm, I = 33 mW/cm^2, D = 2000 mJ/cm^2). The PDL coated patterned foil was glued to the bottom of a 96 well micro titer plate by commercial double sided tape, and used for neuronal cell growth tests.

Rat cortical neurons, (Innoprot Cat N: P10102) at 50 k cells/well cell density were placed on the surfaces and incubated up to 72 h with neurobasal culture medium (05790 Stemcell Technologies, Vancouver, Canada) supplemented with NeuroCult SM1 (05711, Stemcell Technologies). Then cell bodies were immunostained with anti-beta III Tubulin-Alexa 488 antibody (AB15708A4, Sigma-Aldrich, St. Louis, MO, USA) and nuclei with DAPI (D9542, Sigma-Aldrich, St. Louis, MO, USA) following the manufacturer's indications. Images were analyzed by confocal microscopy and processed using Wimasis-Wimneuron software (Onimagin Technologies, Cordoba, Spain).

3. Results and Discussion

3.1. Mastering

As described above. the micro patterned bottom foil of a micro titer plate was designed in a way that a 1000 μm thick feeding channel is placed around a network (see Figure 1) of singular micro cavities (diameter 40 μm) connected to each other by microchannels (length 140 μm, 5 μm wide). Microfluidic simulation showed that the rounded edges on the interconnection between channel and micro cavity enabled a faster and more homogenous filling of the resulting microfluidic network. For the micro patterned bottom in the microcavities, a pattern height of 17 μm was found to be optimal between fast production cycles of the pattern and spatial separation of the bottom of the micro titer plate from the triangular plateaus in order to maximize the optical output and minimize internal interfering signals.

3.2. P2P UV-NIL Imprints

A hard working stamp derived from the orientation A (see Scheme 2) was used to replicate the pattern into the low viscose UV-NIL resist. Details on cytotoxicity and cell compatibility were tested and are discussed in detail later.

The replication of the desired pattern worked well and defect-free imprints were obtained using PMMA as the substrate (see Figure 2). Processing the resist via spin coating resulted in a too thin film (approximately 700 nm at 3000 rpm) for the required pattern height. Hence, other deposition methods had to be tested. As most promising and most comparable to the high-throughput R2R process, a 10 µm thin film was applied via doctor blading. The so obtained film of uncured resist was manually brought into contact with the working stamp. Under a microscope the filling of the microscopic cavities could be observed. It is noteworthy that filling time s comparatively high due to large aspect ratio of the desired structure and the limited occurrence of capillary forces. With such large patterns, the process step of filling becomes increasingly demanding in time as the capillary forces that help filling small features sizes only play a minor role. Further, the volume of air that needs to be removed and replaced by resist in such large micron sized features is significantly larger compared to nano patterns resulting in an even slower process. However, the low viscosity of the material enables a faster absorption of the entrapped air, in comparison to other high viscose resists. Nevertheless, in the case of the tested pattern, filling times of roughly 10 min were observed. One advantage of the R2R based approach, besides the potential high volume production, is the continuous filling and curing during imprinting. With a sufficiently well-designed shim, trapped air can be avoided and "pushed out" by the liquid resist front on the roller [42].

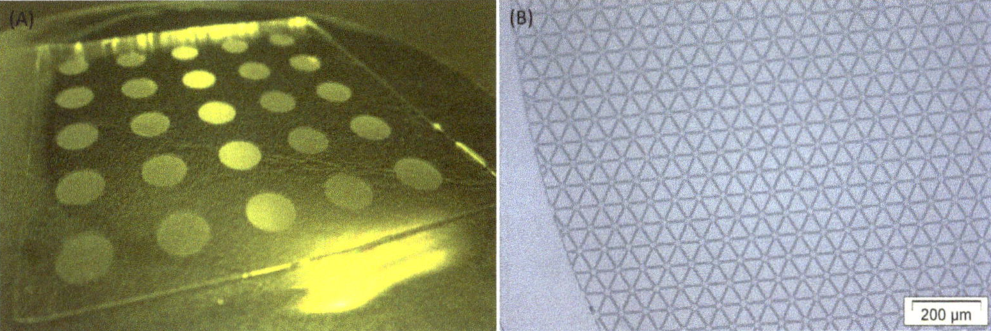

Figure 2. Imprint on PMMA via P2P-imprinting with a hard working stamp. (**A**) Photograph of the imprint and (**B**) microscope image showing the microfluidic network.

The obtained imprints were used to conduct the initial filling tests of the open microfluidic channels via capillary forces. This was a necessary prerequisite for functionalization of the cavities and channels with a neuron growth-enabling substance such as fibronectin. It was observed that with a structure height of 17 µm, spontaneous filling of the channels and cavities with aqueous solution was possible for sufficiently hydrophilic surfaces. In the case of the used UV-NIL material, the low surface tension of the resist prevented spontaneous channel filling. However, it was found that a short oxygen plasma treatment (100 W, 10 s) was sufficient to increase the hydrophilicity of the cured resist and spontaneous channel filling was obtained. Nonetheless, it could be shown that such short plasma was sufficient to keep the selectivity of filling only the channels with aqueous solutions and thus paving the way for subsequent biofunctionalization. This surface hydrophilization was shown to be stable for several weeks after an initial regain in hydrophobicity (measured by advancing water contact angle, see Figure 3). After this increase, the contact angle of around 30° was stable for 3 months (details on the measurement of the dynamic water contact angle can be found in the supporting information). The slow increase can be explained by the high network density preventing a rearrangement of hydrophobic groups to the surface. Even though this trend was only investigated for a few months, it is highly likely that the cured film will never reach its initial hydrophobicity and will stay hydrophilic.

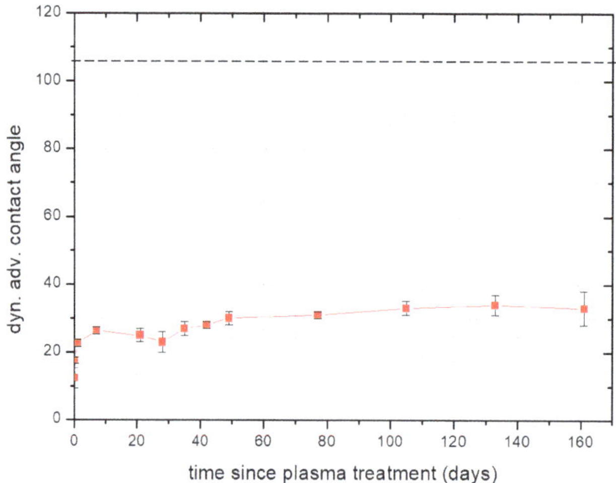

Figure 3. Advancing dynamic water contact angle (WCA) of a cured film after a short oxygen plasma treatment and its development over time. The dashed line indicates the initial adv. dyn. WCA of 107° before plasma treatment.

3.3. Converting to R2R UV-NIL

Roll-to-roll (R2R) imprinting is a continuous fabrication method in which a film of UV-curable material is applied on a flexible substrate and patterned via a roller based shim (see Scheme 3).

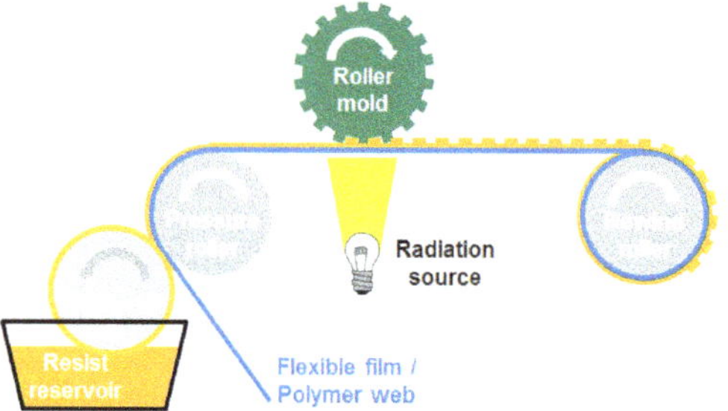

Scheme 3. Schematic representation of the R2R fabrication process.

Similar to the P2P UV-NIL imprints, avoiding the entrapment of air and giving enough time to fill the pattern were crucial to obtain a defect free imprint. Hence, the comparatively slow roller speeds of only up to 2 m/min were applied and visual inspection under a light microscope of the imprinted foils suggested a defect free imprint at speeds below 1 m/min. As seen in Figure 4 and Figure S1, with an increasing web speed the amount of air defects increased. At 0.5 m/min almost no entrapped air was visible, while at 2 m/min a high defect rate could be observed. The entrapped air bubbles were placed on the last corner of the triangles, where they were pushed together. With increasing web speed, the amount of entrapped air increased, as shown in the microscope images. In comparison to other less demanding patterns, the chosen design did not allow a complete pushing out of the air and only at speeds up to 1 m/min was the absorption of the entrapped air fast enough for a defect free imprint. It is noteworthy that other resists with higher viscosity were tested as well, but no defect free imprints were obtained. With the optimized shim design, patterned foils for around 1200 full micro titer plates can be imprinted in one hour showing the potential for mass fabrication.

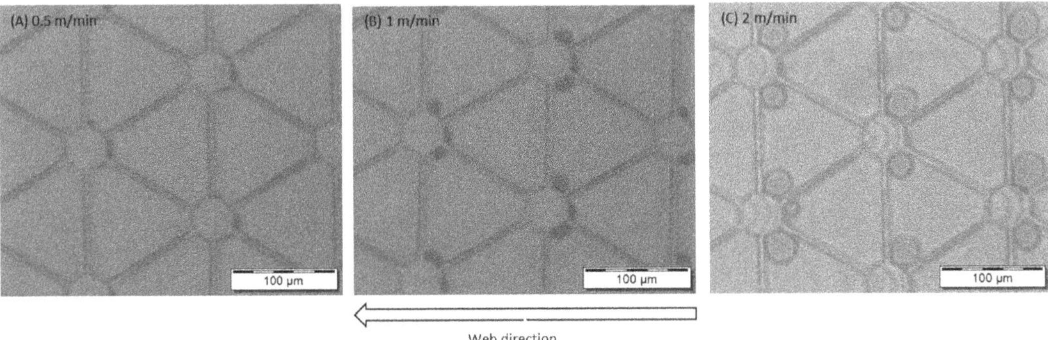

Figure 4. Microscope images of the patterned surface after R2R UV-NIL imprinting showing the defect rate depending on the imprint speed ((**A**) 0.5 m/min; (**B**) 1 m/min; (**C**) 2 m/min).

3.4. Cytotoxicity Validation

For a device to be used in contact with cells or neurons, it needs to be proven to be non-cytotoxic. Following ISO10993-2005 mr-UVCur26SF is to be considered non-cytotoxic under the mentioned curing conditions. However, since the resist is a photocurable material and the used photoinitiator shows a dependence on wavelength and intensity, the curing properties might alter depending on the given fabrication method. Further, impurities resulting from a rub off of the stamp used (in the case of the R2R fabrication, an aluminum coating was used), might interact with the cell lines and create a cytotoxic behavior of the cured surface. We tested an assay of different batches of the resist, as well as different curing conditions (see Table 1) to demonstrate the broad manufacturing window of our material.

Table 1. Summary of different tested curing conditions and the resulting nontoxicity (following UNE-EN-ISO 10993-5: 2009) of mr-UVCur26SF. Standard deviation for all samples was below 5%.

Light Source	λ (nm)	Intensity I (mW/cm^2)	Dose D (mJ/cm^2)	Cell Viability (%)
Hg	UV	10	2000	97 ± 4
Hg	UV	20	500	101 ± 5
Hg	UV	20	1000	99 ± 5
Hg	UV	20	2000	95 ± 3
LED	365	10	2000	93 ± 4
LED	365	20	500	96 ± 4
LED	365	20	1000	91 ± 6
LED	365	20	2000	96 ± 4
LED	365	30	2000	91 ± 4
LED	365	50	2000	91 ± 4
LED	390	20	2000	97 ± 4
LED (R2R) [1]	395	-	-	93 ± 5

[1] Applied roller speed 1 m min^{-1}.

It can be seen that the resist can be cured either by broad band mercury irradiation as well as specific LED wavelength. Further, the intensity can be varied in a range between 10–50 mW/cm^2 and the dose can be optimized between 500 and 2000 mJ/cm^2. Within this broad range of curing conditions none of the samples showed cytotoxic behavior leaving a large process window in which the pattern can be fabricated.

Further, the compatibility of the tested material with different cell lines: kidney tubular epithelial cells (KTEC), conjunctive epithelial cells (Epit. Conj.), hepatic endothelial cells (Hepatic. Endo.), dermal endothelial cells (Skin. Endo.), bone osteosarcoma epithelial cells (U2OS) and embryonic kidney 293 cells (HEK293) has been tested, both on the pristine cured resist and in combination with different biocoatings (lysine and collagen). The cell lines were chosen to show the broad applicability of the mr-UVCur26SF. The cell lines in the investigation contained both tumoral cells (U2OS, HEK) and primary cells. Within the primary cells, they included epithelial and endothelial cells. In addition, there were cells from different sources, monkey, mouse and human. The results were compared to pristine PS, a standard material used for micro titer plates in state of the art assays. Any sample with a viability of 80% or higher was considered as compatible with the cell line. The results are summarized in a qualitative fashion in Figure 5 and Figure S2. A more detailed quantitative analysis is given in the supporting information. Most cell lines showed an overall good compliance with the material. After 72 h, all showed a cell viability that was comparable to state-of-the-art cell assay materials when the surface was treated with short oxygen plasma. Poly-D-lysin showed the best performance of the two tested biocoatings and it could be seen that besides KTEC, all cell lines were compatible with a lysine coating. Collagen on the other hand showed some significant differences in cell viability when used directly or in combination with initial plasma. A clear image for these findings is still under investigation. However, it appears that one major reason might be the polarity of the resist after plasma treatment and resulting rearrangements of the biocoating. Further, it is known that the hardness of the substrate can have an influence on the adhesion strengths of different cells towards the substrate. Both material's properties can influence the compatibility of the resist towards the given cell lines.

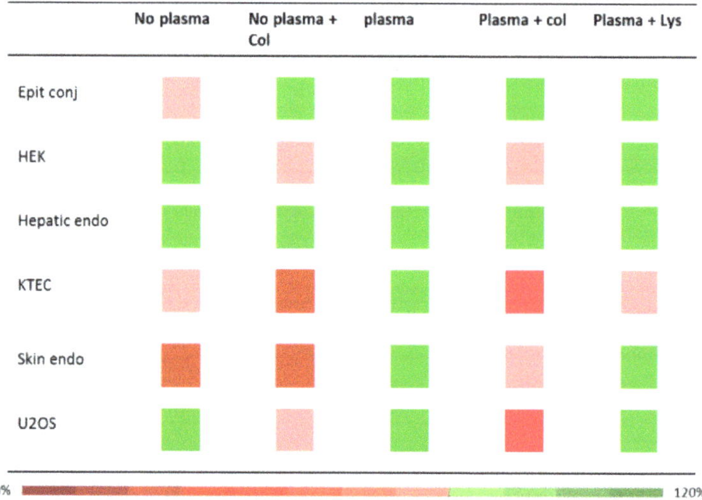

Figure 5. Summary of the cell viability tests of different cell lines after 72 h on a flat surface of mr-UVCur26SF under different coating conditions. The color shift from red to green indicate the increasing viability, with 80% (light green) being sufficient for further testing.

The results shown should function as an indicator to which cell types the mr-UVCur26-SF is compatible. It is noteworthy that an incompatibility towards a specific cell lines does not automatically indicate a toxicity. The given test assay rather indicates the tendency of the cells to adhere on the resist surface. For further tests, a surface functionalization with a short plasma treatment, followed by poly-D-lysine (PDL) treatment was chosen.

3.5. Neuronal Growth

The behavior of neurons in contact to the patterned surfaces was the main objective of these investigations and the compatibility was tested accordingly. The patterned samples were compared to standard unpatterned polystyrene 96 MTP covered with PDL. As can be seen in Figure 6, the neuronal cells aligned well along the channels of the microfluidic system (left panel) and were much better separated in comparison to the flat surface of the standard MTP (right panel). The majority of the neurons were placed within the microfluidic network as the plateaus were not covered with PDL and hence were less adhering for the cells (data not shown). Moreover, using confocal microscopy only, the fluorescent signal from the cells within the patterned structures could be monitored. The selectivity of the cells to attach in the coated areas gave an increased control over the separation, allowing a more organized culture. The overall number of neuronal cells decreased in comparison with the flat PS surface. This simplified the further imaging analysis. Via confocal fluorescence microscopy, single cells could be observed and their neurite outgrowth studied. With these promising results neuroprotection assays have been conducted to test the improvement of our approach over current protocols.

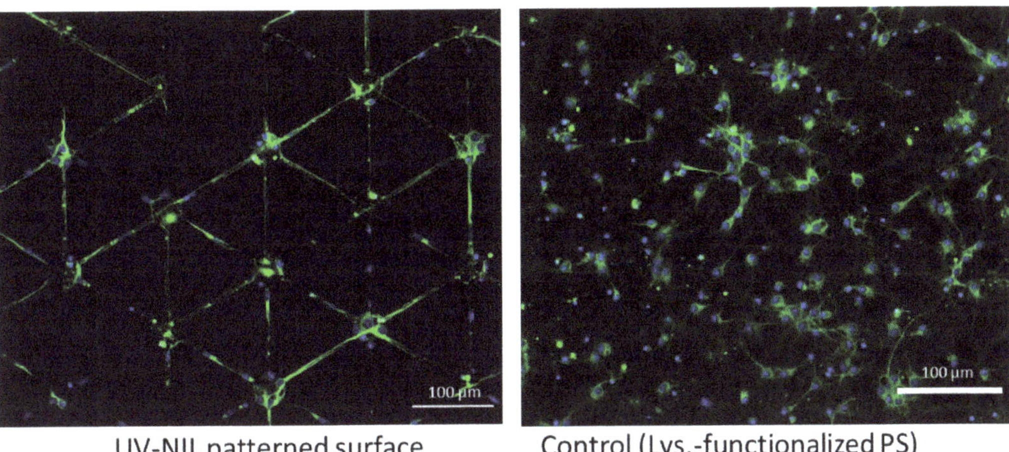

Figure 6. Confocal microscope image of seeded neurons on a patterned surface of the used UV-NIL material (**left**) and an unstructured PS micro titer plate (**right**) for reference. Neurites (cell bodies are stained with Anti Tubulin III-488 antibody (green) and nuclei with DAPI (blue)) aligned along the microfluidic channel.

For such assay, neuronal cells were cultured in neurobasal growth medium in the absence and presence of a neuroprotective compound, MK-801 (Catalog number M-107, Sigma-Aldrich, St. Louis, MO, USA) at 10 µM concentration. Then, 100 µM glutamate solution was added to provoke injury to cells. After 15 min the glutamate was replaced by fresh medium and thoroughly washed. 48 h later, cells were immunostained with Anti-Tubulin III-488 antibody and imaged on a confocal microscope. Control wells were also prepared with neither glutamate nor MK-801 incubation. Different parameters of neuronal growth, such as the total number of vital neurons, length of neurites, total branching points or others were studied. The data shown in Figure 7 and Table S1 are the average of three independent studies, in which individual wells were seated with the same number of cells, treated in the same way and quantified individually.

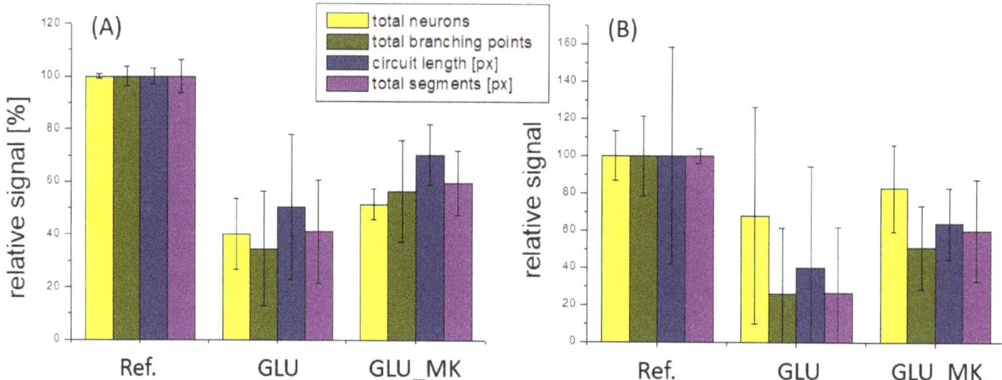

Figure 7. Influence of a neurotoxin (glutamate, GLU) on the total neuron count, total branching points, circuit length and total segments for a patterned surface (**A**) and a flat PS surface (**B**). Additionally, the results with a neuron protective agent (MK-801, GLU_MK) are shown.

The results indicate that neurons react to the external stimulus of glutamate and MK-801 similarly, both on the flat PS surface (Figure 7B) and on the patterned surface (Figure 7A). In contact with the neuronal cell toxin (glutamate), the number of viable neurons, their branching points, circuit length and total segments decreased, as expected. This loss in viability can be partially overcome using a neuron protective (MK-801). While it is difficult to compare the results from the flat PS surface with the one patterned with a UV-NIL material, as essential parameters such as total cell count and cell density differ dramatically, a few crucial findings can be made. Firstly, the overall trend in both cases in terms of damage and recovery of the neurons was similar when using patterned surfaces in comparison with standard plate protocol. A decrease in viability was observable when the neurons came in contact with the toxin and a certain protection from this toxin could be achieved using an appropriate reagent. This indicates that the neurons within the trenches stay in contact with the medium above the micro grooves and reagents can migrate towards the cells. This can be considered as an improvement because it allows the use of lower cell numbers for the same result. Secondly, the conformal stress that was put on the cells, needing to adjust themselves towards the pattern, did not result in a higher sensitivity for neuronal toxins. Thirdly, and most importantly, the reduced number of cells, and thereby easier automated "assignment" to specific cells led to a reduced deviation in all investigated parameters. The total number of cells on the patterned surface was only approximately a third in comparison to the PS reference. However, the deviation in cell numbers was reduced from 13% to 0.8%. The branching points decreased by a factor of approximately 6.5 between the patterned and unpatterned surfaces, while the deviation was 3.5% in comparison to 21.6%. The total circuitry length was about 2.2 times shorter than on the flat surface, but the deviation was 2.7% in comparison to 61% of the unpatterned surface. For the total segment length decreased by a factor of roughly 5.6, the deviation stayed almost the same for both cases. Similar results could be observed for the samples treated with the neurotoxin and the neuron protective agent. In the supporting information the absolute values are given from which the relative numbers were derived. While we are aware that this assay can only function as a proof-of-principle and more data need to be created to reduce the deviation shown in the results, we believe that the pattern we suggest can help to create more accurate automated read-out systems by reducing the total neuron count in a defined area and thereby also reducing the entanglement. Given this, better reliability is expected between experiments and therefore the analysis would be more robust. Further, it can be seen that a standard protocol for analyzing the behavior of neurons towards specific drugs can be easily adopted towards the patterned surface without significant changes or needs for optimization. In addition, this organized arrangement will allow other types of applications and experiments apart from neurite outgrowth or even with other types of cells.

4. Conclusions

In this paper, we have shown a straight forward approach to decrease the standard deviation of a neuronal cell assay by separating cell clusters with a micro patterned surface of a standard micro titer plate for easy read-out. Using industrial compatible processes, we have fabricated polymer shims and working stamps to manufacture such patterned surfaces both in a small scale P2P fashion as well as continuous high throughput R2R fabrication. Furthermore, we have evaluated the cytotoxicity and cell compatibility of our chosen material, the fast curing mr-UVCur26SF, for a multitude of different cell lines and curing conditions, showing the high compatibility of the material, even beyond the scope of the publication. Finally, initial neuronal cell assay tests were conducted, comparing the chosen patterned surface with a flat polystyrene reference, showing that our initial idea of spatial separation of neurons prevents clustering to some extent and reduces the overall standard deviation of the preformed assay. We believe that with our approach a new class of neuron based drug screening assays can be developed, leading to faster and more reliable high-throughput drug screening.

Supplementary Materials: The following are available online at https://www.mdpi.com/article/10.3390/nano11040902/s1, Figure S1: comparison of a low viscous (15 mPas) UV-NIL resist with a higher viscous, Figure S2: unpatterned surface of mr-UVCur26SF: (a) Pristine; (b) collagene coated on pristine surface; (c) O_2 plasma treated surface (d) collagene coated on O_2 plasma treated surface (e) poly-D-lysine coated on O_2 plasma treated surface, Table S1: Average values of two separate sets of experiments with the given deviation in their results.

Author Contributions: Conceptualization, M.L. and M.W.T.; funding acquisition, C.S. and A.S.; investigation, M.L., A.H., N.B.I., A.A.I. and I.R.; project administration, M.S.; writing—original draft, M.L.; writing—review and editing, M.W.T., A.H., N.B.I., I.R. and A.S. All authors have read and agreed to the published version of the manuscript.

Funding: This project has received funding from the European Union's Horizon 2020 research and innovation programme under grant agreement NO. 646260. The research was also partially supported by NextGenMicrofluidics project under HORIZON2020 with grant agreement no 862092.

Institutional Review Board Statement: Not applicable.

Informed Consent Statement: Not applicable.

Acknowledgments: The authors thank Anett Kolander, Ludwig Scharfenberg and Maren Schiersch for their valuable assistance of process engineering.

Conflicts of Interest: The authors declare no conflict of interest.

References

1. Deleglise, B.; Lassus, B.; Soubeyre, V.; Alleaume-Butaux, A.; Hjorth, J.J.; Vignes, M.; Schneider, B.; Brugg, B.; Viovy, J.-L.; Peyrin, J.-M. Synapto-Protective Drugs Evaluation in Reconstructed Neuronal Network. *PLoS ONE* **2013**, *8*, e71103. [CrossRef]
2. Larsson, P.; Engqvist, H.; Biermann, J.; Rönnerman, E.W.; Forssell-Aronsson, E.; Kovács, A.; Karlsson, P.; Helou, K.; Parris, T.Z. Optimization of cell viability assays to improve replicability and reproducibility of cancer drug sensitivity screens. *Sci. Rep.* **2020**, *10*, 5798. [CrossRef] [PubMed]
3. Nierode, G.; Kwon, P.S.; Dordick, J.S.; Kwon, S.-J. Cell-Based Assay Design for High-Content Screening of Drug Candidates. *J. Microbiol. Biotechnol.* **2016**, *26*, 213–225. [CrossRef] [PubMed]
4. Kepp, O.; Galluzzi, L.; Lipinski, M.; Yuan, J.; Kroemer, G. Cell death assays for drug discovery. *Nat. Rev. Drug Discov.* **2011**, *10*, 221–237. [CrossRef] [PubMed]
5. Zang, R.; Li, D.; Tang, I.C.; Wang, J.; Yang, S.T. Cell-Based Assays in High-Throughput Screening for Drug Discovery. *Int. J. Biotechnol. Wellness Ind.* **2012**, *1*, 31–51. [CrossRef]
6. Johnston, P.A. Redox cycling compounds generate H2O2 in HTS buffers containing strong reducing reagents—Real hits or promiscuous artifacts? *Curr. Opin. Chem. Biol.* **2011**, *15*, 174–182. [CrossRef] [PubMed]

7. Drewe, J.; Kasibhatla, S.; Tseng, B.; Shelton, E.; Sperandio, D.; Yee, R.M.; Litvak, J.; Sendzik, M.; Spencer, J.R.; Cai, S.X. Discovery of 5-(4-hydroxy-6-methyl-2-oxo-2H-pyran-3-yl)-7-phenyl-(E)-2,3,6,7-tetrahydro-1,4-thiazepines as a new series of apoptosis inducers using a cell- and caspase-based HTS assay. *Bioorg. Med. Chem. Lett.* **2007**, *17*, 4987–4990. [CrossRef] [PubMed]
8. Peuhu, E.; Paul, P.; Remes, M.; Holmbom, T.; Eklund, P.; Sjöholm, R.; Eriksson, J.E. The antitumor lignan Nortrachelogenin sensitizes prostate cancer cells to TRAIL-induced cell death by inhibition of the Akt pathway and growth factor signaling. *Biochem. Pharmacol.* **2013**, *86*, 571–583. [CrossRef] [PubMed]
9. Shiloh, Y.; Andegeko, Y.; Tsarfaty, I. In search of drug treatment for genetic defects in the DNA damage response: The example of ataxia-telangiectasia. *Semin. Cancer Biol.* **2004**, *14*, 295–305. [CrossRef]
10. Yeyeodu, S.T.; Witherspoon, S.M.; Gilyazova, N.; Ibeanu, G.C. A Rapid, Inexpensive High Throughput Screen Method for Neurite Outgrowth. *Curr. Chem. Genom.* **2010**, *4*, 74–83. [CrossRef] [PubMed]
11. Roach, P.; Parker, T.; Gadegaard, N.; Alexander, M. Surface strategies for control of neuronal cell adhesion: A review. *Surf. Sci. Rep.* **2010**, *65*, 145–173. [CrossRef]
12. Hu, J.; Hardy, C.; Chen, C.-M.; Yang, S.; Voloshin, A.S.; Liu, Y. Enhanced Cell Adhesion and Alignment on Micro-Wavy Patterned Surfaces. *PLoS ONE* **2014**, *9*, e104502. [CrossRef]
13. Kusuma, S.; Smith, Q.; Facklam, A.; Gerecht, S. Micropattern size-dependent endothelial differentiation from a human induced pluripotent stem cell line. *J. Tissue Eng. Regen. Med.* **2017**, *11*, 855–861. [CrossRef] [PubMed]
14. Joo, S.; Kim, J.Y.; Lee, E.; Hong, N.; Sun, W.; Nam, Y. Effects of ECM protein micropatterns on the migration and differentiation of adult neural stem cells. *Sci. Rep.* **2015**, *5*, 13043. [CrossRef]
15. Buzanska, L.; Zychowicz, M.; Ruiz, A.; Rossi, F. *Stem Cell Technologies in Neuroscience*; Neuromethods; Humana Press: New York, NY, USA, 2017; Volume 126, pp. 19–44. [CrossRef]
16. Lücker, P.B.; Javaherian, S.; Soleas, J.P.; Halverson, D.; Zandstra, P.W.; McGuigan, A.P. A microgroove patterned multiwell cell culture plate for high-throughput studies of cell alignment. *Biotechnol. Bioeng.* **2014**, *111*, 2537–2548. [CrossRef]
17. Roach, P.; Parker, T.; Gadegaard, N.; Alexander, M.R. A bio-inspired neural environment to control neurons comprising radial glia, substrate chemistry and topography. *Biomater. Sci.* **2013**, *1*, 83–93. [CrossRef] [PubMed]
18. Fendler, C.; Harberts, J.; Rafeldt, L.; Loers, G.; Zierold, R.; Blick, R.H. Neurite guidance and neuro-caging on steps and grooves in 2.5 dimensions. *Nanoscale Adv.* **2020**, *2*, 5192–5200. [CrossRef]
19. Li, W.; Tang, Q.Y.; Jadhav, A.D.; Narang, A.; Qian, W.X.; Shi, P.; Pang, S.W. Large-scale Topographical Screen for Investigation of Physical Neural-Guidance Cues. *Sci. Rep.* **2015**, *5*, 8644. [CrossRef]
20. Marcus, M.; Baranes, K.; Park, M.; Choi, I.S.; Kang, K.; Shefi, O. Interactions of Neurons with Physical Environments. *Adv. Healthc. Mater.* **2017**, *6*, 1700267. [CrossRef]
21. Dalby, M.J.; Gadegaard, N.; Tare, R.; Andar, A.; Riehle, M.O.; Herzyk, P.; Wilkinson, C.D.W.; Oreffo, R.O.C. The control of human mesenchymal cell differentiation using nanoscale symmetry and disorder. *Nat. Mater.* **2007**, *6*, 997–1003. [CrossRef]
22. Reynolds, P.M.; Pedersen, R.H.; Stormonth-Darling, J.; Dalby, M.J.; Riehle, M.O.; Gadegaard, N. Label-Free Segmentation of Co-cultured Cells on a Nanotopographical Gradient. *Nano Lett.* **2013**, *13*, 570–576. [CrossRef] [PubMed]
23. Yang, J.; McNamara, L.E.; Gadegaard, N.; Alakpa, E.V.; Burgess, K.V.; Meek, R.M.D.; Dalby, M.J. Nanotopographical Induction of Osteogenesis through Adhesion, Bone Morphogenic Protein Cosignaling, and Regulation of MicroRNAs. *ACS Nano* **2014**, *8*, 9941–9953. [CrossRef]
24. Ahfeldt, T.; Litterman, N.K.; Rubin, L.L. Studying human disease using human neurons. *Brain Res.* **2017**, *1656*, 40–48. [CrossRef] [PubMed]
25. Gladkov, A.; Pigareva, Y.; Kutyina, D.; Kolpakov, V.; Bukatin, A.; Mukhina, I.; Kazantsev, V.; Pimashkin, A. Design of Cultured Neuron Networks in vitro with Predefined Connectivity Using Asymmetric Microfluidic Channels. *Sci. Rep.* **2017**, *7*, 15625. [CrossRef] [PubMed]
26. Forró, C.; Thompson-Steckel, G.; Weaver, S.; Weydert, S.; Ihle, S.; Dermutz, H.; Aebersold, M.J.; Pilz, R.; Demkó, L.; Vörös, J. Modular microstructure design to build neuronal networks of defined functional connectivity. *Biosens. Bioelectron.* **2018**, *122*, 75–87. [CrossRef] [PubMed]
27. Peyrin, J.-M.; Deleglise, B.; Saias, L.; Vignes, M.; Gougis, P.; Magnifico, S.; Betuing, S.; Pietri, M.; Caboche, J.; Vanhoutte, P.; et al. Axon diodes for the reconstruction of oriented neuronal networks in microfluidic chambers. *Lab Chip* **2011**, *11*, 3663–3673. [CrossRef] [PubMed]
28. Lee, N.; Park, J.W.; Kim, H.J.; Yeon, J.H.; Kwon, J.; Ko, J.J.; Oh, S.-H.; Kim, H.S.; Kim, A.; Han, B.S.; et al. Monitoring the differentiation and migration patterns of neural cells derived from human embryonic stem cells using a microfluidic culture system. *Mol. Cells* **2014**, *37*, 497–502. [CrossRef]
29. Kim, Y.-T.; Karthikeyan, K.; Chirvi, S.; Dave, D.P. Neuro-optical microfluidic platform to study injury and regeneration of single axons. *Lab Chip* **2009**, *9*, 2576–2581. [CrossRef]
30. Maisonneuve, B.G.C.; Vieira, J.; Larramendy, F.; Honegger, T. Microchannel patterning strategies for in vitro structural connectivity modulation of neural networks. *bioRxiv* **2021**. [CrossRef]
31. Yamada, A.; Vignes, M.; Bureau, C.; Mamane, A.; Venzac, B.; Descroix, S.; Viovy, J.-L.; Villard, C.; Peyrin, J.-M.; Malaquin, L. In-mold patterning and actionable axo-somatic compartmentalization for on-chip neuron culture. *Lab Chip* **2016**, *16*, 2059–2068. [CrossRef]

32. Toren, P.; Smolka, M.; Haase, A.; Palfinger, U.; Nees, D.; Ruttloff, S.; Kuna, L.; Schaude, C.; Jauk, S.; Rumpler, M.; et al. High-throughput roll-to-roll production of polymer biochips for multiplexed DNA detection in point-of-care diagnostics. *Lab Chip* **2020**, *20*, 4106–4117. [CrossRef] [PubMed]
33. Thesen, M.W.; Nees, D.; Ruttloff, S.; Rumler, M.; Rommel, M.; Schlachter, F.; Grützner, S.; Vogler, M.; Schleunitz, A.; Grützner, G. Inkjetable and photo-curable resists for large-area and high-throughput roll-to-roll nanoimprint lithography. *J. Micro/Nano Lithogr. MEMS MOEMS* **2014**, *13*, 43003. [CrossRef]
34. Lohse, M.; Heinrich, M.; Grützner, S.; Haase, A.; Ramos, I.; Salado, C.; Thesen, M.W.; Grützner, G. Versatile fabrication method for multiscale hierarchical structured polymer masters using a combination of photo- and nanoimprint lithography. *Micro Nano Eng.* **2021**, *21*, 100079. [CrossRef]
35. Schift, H.; Saxer, S.; Park, S.; Padeste, C.; Pieles, U.; Gobrecht, J. Controlled co-evaporation of silanes for nanoimprint stamps. *Nanotechnology* **2005**, *16*, S171–S175. [CrossRef]
36. Schleunitz, A.; Vogler, M.; Fernandez-Cuesta, I.; Schift, H.; Gruetzner, G. Innovative and Tailor-made Resist and Working Stamp Materials for Advancing NIL-based Production Technology. *J. Photopolym. Sci. Technol.* **2013**, *26*, 119–124. [CrossRef]
37. Decrop, D.; Pardon, G.; Brancato, L.; Kil, D.; Shafagh, R.Z.; Kokalj, T.; Haraldsson, T.; Puers, R.; Van Der Wijngaart, W.; Lammertyn, J. Single-Step Imprinting of Femtoliter Microwell Arrays Allows Digital Bioassays with Attomolar Limit of Detection. *ACS Appl. Mater. Interfaces* **2017**, *9*, 10418–10426. [CrossRef] [PubMed]
38. Lee, H.; Koh, D.; Xu, L.; Row, S.; Andreadis, S.T.; Oh, K.W. A Simple Method for Fabrication of Microstructures Using a PDMS Stamp. *Micromachines* **2016**, *7*, 173. [CrossRef]
39. Vajrala, V.S.; Suraniti, E.; Rigoulet, M.; Devin, A.; Sojic, N.; Arbault, S. PDMS microwells for multi-parametric monitoring of single mitochondria on a large scale: A study of their individual membrane potential and endogenous NADH. *Integr. Biol.* **2016**, *8*, 836–843. [CrossRef]
40. Okulova, N. *High Speed Fabrication of Large-Scale Micro-Patternedthermoplastic Films for Industrial Applications NNT 2018*; NNT: Braga, Portugal, 2018.
41. Lee, J.; Bong, J.; Ha, Y.-G.; Park, S.; Ju, S. Durability of self-assembled monolayers on aluminum oxide surface for determining surface wettability. *Appl. Surf. Sci.* **2015**, *330*, 445–448. [CrossRef]
42. Peng, L.; Yi, P.; Wu, H.; Lai, X. Study on bubble defects in roll-to-roll UV imprinting process for micropyramid arrays II: Numerical study. *J. Vac. Sci. Technol. B* **2016**, *34*, 051203. [CrossRef]

 *nanomaterials*

Article

Mastering of NIL Stamps with Undercut T-Shaped Features from Single Layer to Multilayer Stamps

Philipp Taus [1], Adrian Prinz [2], Heinz D. Wanzenboeck [1,*], Patrick Schuller [1], Anton Tsenov [1], Markus Schinnerl [1], Mostafa M. Shawrav [3], Michael Haslinger [4] and Michael Muehlberger [4]

[1] TU Wien, Institute for Solid State Electronics, 1040 Vienna, Austria; philipp.taus@tuwien.ac.at (P.T.); patrick.schuller@tuwien.ac.at (P.S.); anton.tsenov@tuwien.ac.at (A.T.); markus.schinnerl@tuwien.ac.at (M.S.)
[2] Stratec Consumables GmbH, 5081 Anif, Austria; a.prinz@stratec.com
[3] ams AG, Tobelbader Strasse 30, 8141 Premstaetten, Austria; mostafa.shawrav@gmail.com
[4] PROFACTOR GmbH, 4407 Steyr, Austria; michael.haslinger@profactor.at (M.H.); michael.muehlberger@profactor.at (M.M.)
* Correspondence: heinz.wanzenboeck@tuwien.ac.at; Tel.: +43-1-588-36243

Abstract: Biomimetic structures such as structural colors demand a fabrication technology of complex three-dimensional nanostructures on large areas. Nanoimprint lithography (NIL) is capable of large area replication of three-dimensional structures, but the master stamp fabrication is often a bottleneck. We have demonstrated different approaches allowing for the generation of sophisticated undercut T-shaped masters for NIL replication. With a layer-stack of phase transition material (PTM) on poly-Si, we have demonstrated the successful fabrication of a single layer undercut T-shaped structure. With a multilayer-stack of silicon oxide on silicon, we have shown the successful fabrication of a multilayer undercut T-shaped structures. For patterning optical lithography, electron beam lithography and nanoimprint lithography have been compared and have yielded structures from 10 µm down to 300 nm. The multilayer undercut T-shaped structures closely resemble the geometry of the surface of a Morpho butterfly, and may be used in future to replicate structural colors on artificial surfaces.

Keywords: nanoimprint lithography (NIL); undercut features; master; Blu-Ray patterning; reactive ion etching

Citation: Taus, P.; Prinz, A.; Wanzenboeck, H.D.; Schuller, P.; Tsenov, A.; Schinnerl, M.; Shawrav, M.M.; Haslinger, M.; Muehlberger, M. Mastering of NIL Stamps with Undercut T-Shaped Features from Single Layer to Multilayer Stamps. *Nanomaterials* **2021**, *11*, 956. https://doi.org/10.3390/nano11040956

Academic Editor: Konstantins Jefimovs

Received: 25 February 2021
Accepted: 19 March 2021
Published: 9 April 2021

Publisher's Note: MDPI stays neutral with regard to jurisdictional claims in published maps and institutional affiliations.

1. Introduction

In biology, surfaces with three-dimensional nanostructures are responsible for outstanding physicochemical properties of the surface. The anti-wetting surface of the lotus flower, the low friction surface of shark skin and the high sticking capability of Gecko feet are well-described. Another exceptional example is structural color: Not chemical pigmentation, but a complex structured multilayered nanostructure, is the cause of the brilliant blue wing colors in some of the species of morpho butterflies and the golden color of the beetle *Chrysina aurigans* [1,2].

Imitating nature's nanostructured surfaces by engineering methods promises to gain the same spectacular properties on human-made products. The biomimetic replication of a lotus-like surface for anti-wetting features has been described [3]. A wide range of lithographic methods has been successfully applied for three-dimensionally structured surfaces [4–7]. However, many methods can only generate these biomimetic structures on a small area. The challenge is the fabrication on a larger surface area—here nanoimprint lithography has a lot of potential [8].

Nanoimprint lithography (NIL) is a cost-effective method to replicate nanostructures on large areas [9,10]. A nanostructured stamp is used to mechanically emboss a polymer layer on a substrate [11]. Such three-dimensionally structured surfaces are interesting for microfluidic channels [12,13], superhydrophobic surfaces [14], and optical applications [15,16].

Also undercut structures [11,17] and hierarchical structures [18] can be fabricated. With an imprint stepper or with roll-to-roll NIL also large areas can be patterned three-dimensionally. For fabrication of three-dimensional structures hybrid structuring processes combining NIL and electron beam lithography [19,20] or NIL and photolithography [21], as well as multilayer nanoimprinting [22] to create hierarchical stamp masters for optical micro- and nanostructures have been recently demonstrated successfully. Even imprinting of complex three-dimensional nanomolds [23] was successfully demonstrated.

In contrast to the layer by layer fabrication with conventional multilevel lithography, the NIL fabrication allows to directly replicate a complete 3D structure in a single patterning step [24]. Master stamps are a critical component in the nanoimprint process. While initially these masters were directly used as stamps, now it is more common to replicate a working stamp from the original master and then use this working stamp for the actual nanoimprint process of the final product.

Often the masters are made from silicon wafers using well-established silicon technology, like electron beam lithography or interference lithography combined with reactive ion etching [10], but also biological samples can be used to directly replicate bionic structures [4,25]. The fabrication of three-dimensional master stamps has also been achieved by focused ion beam related methods [26,27] or by grey scale lithography [28].

The replication of nanostructures in the nanoimprint process is commonly limited to draft angles equal or smaller than $90°$, but recently it was also shown that undercut structures as well as nano-cavities could be replicated when the right materials for the stamp and imprint material are chosen [11,29]. The process parameters of the replication—especially of the demolding step—are critical for freeform microstructures [30]. Also, innovative alternatives such as shape memory polymers have been reported [31,32]. Not only is the replication process a challenge, but also the fabrication of the master itself.

The objective of this work is to fabricate the multilevel nanostructured surface resembling geometry of a Morpho menelaus wing. The challenge is that multilayer undercut features have to be fabricated. Such structures are not only beneficial for biomimetic structures, a similar requirement also exists for vertical-channel solid-state memory [33,34]. This paper describes and compares the fabrication of (A) single-layer undercut T-structures using PTM-layers (phase transition material) on silicon as well as (B) multi-layer undercut T-structures of a Si-SiO$_2$-multi-stack (Table 1). The PTM-layers on silicon were patterned mask-less by Blu-Ray laser writing [35]. The Si-SiO$_2$-multi-stack was etched by RIE using gold hardmasks patterned by three different lithography methods—optical lithography, electron beam lithography, and NIL.

Table 1. Overview of five process sequences used to fabricate undercut T-structures

<table>
<tr><th>T-Structure</th><th colspan="2">Single-Layer</th><th colspan="3">Multi-Layer</th></tr>
<tr><td>Patterning</td><td colspan="2">Blu-Ray laser</td><td>g-line lithography</td><td>e-beam lithography</td><td>NIL</td></tr>
<tr><td>Lift-off mask</td><td colspan="2">-</td><td>AZ5214E</td><td>PMMA</td><td>mr-NIL212</td></tr>
<tr><td>Etch-mask</td><td colspan="2">Inorganic PTM-layer</td><td colspan="3">Au-hardmask</td></tr>
<tr><td>Etch Step 1</td><td rowspan="2">KOH wet etch</td><td rowspan="2">Medium pressure RIE</td><td colspan="3">Low pressure RIE</td></tr>
<tr><td>Etch Step 2</td><td colspan="3">High pressure RIE</td></tr>
</table>

2. Materials and Methods

As substrate for the master, Si (100) wafers were used. These wafers were used to fabricate two types of masters: (i) single-layer undercut structures with a T-shaped cross-section and (ii) multilayer undercut structures with a T-on-top-of-T-shaped cross-section. To accomplish these structures, a multilayer stack of different materials was deposited by sputter deposition. The materials deposited were silicon, silicon oxide and phase transition material (PTM) [36]. The lateral structure was defined by lithographic patterning of the top layer as described in the following sections:

2.1. Single Layer Undercut Masters

To prepare the single-layer T-shaped masters (see Figure 1) a layer stack of PTM (MoWO$_x$), amorphous silicon (a-Si) and PTM were deposited on a 200 mm Si wafer by sputter deposition. The top PTM layer was 50 nm thick and was processed to act as a hard mask. The bottom PTM layer served as an etch-stop layer. For patterning of the top layer, the PTR-3000 phase transition mastering system (Sony Disc Technology Inc., Tokyo, Japan) was used. The PTR 3000 system was developed for Blu-Ray mastering and uses a 405 nm diode laser for exposure [37]. Blu-Ray mastering allows high-speed large area writing with resolutions down to 130 nm [35]. The phase transition material behaves comparable to a positive photoresist and changes from amorphous to polycrystalline upon laser irradiation. Compared to regular Blu-Ray mastering, this process used constant angular velocity writing with a rotation velocity of 450 rpm and a 7.75 ns clock cycle, which results in pit sizes of 7.3 nm at 20 mm radius and 14.6 nm at 40 mm. The pit-density was increased from inner tracks towards outer tracks (trackpitch 40 nm) in software, to achieve a homogenous exposure. The exposed area can be developed using a 2.38% TMAH wet chemistry for 2000 s. With the top PTM layer as hard mask, the underlying layer of amorphous Si was wet etched using 25% KOH solution at 40 °C. As the amorphous Si lacks any crystalline order, the etching by KOH is isotropic and also etches sideways. The geometry of the undercut features was dependent on the duration of this isotropic etch process and was varied from 2–10 min.

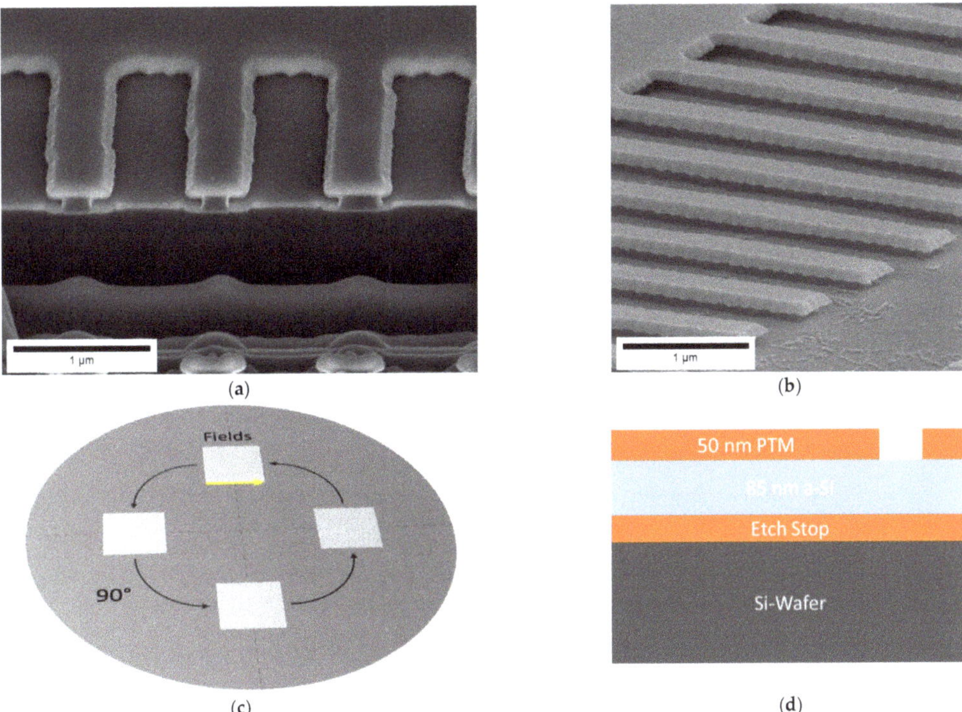

Figure 1. Single layer undercut master. (**a**) SEM image of a FIB-cross-section showing the underetched geometry (**b**) Tilted view SEM image of the single-layer underetched structures. The PTM top layer forms the overhanging ledge on top of the underetched Si base. (**c**) Optical image of wafer is indicating the Blu-Ray mastering of the top PTM layer (**d**) Schematic cross-section through the layer stack.

Alternatively, also a reactive ion etching (RIE) process was developed to avoid wet chemical etching. An RF power setting of 30 W with a medium chamber pressure of 10 mTorr, a gas flow of 20 sccm SF_6 and 10 sccm Ar was used. The etch rate in Si was at least one magnitude faster than in the PTM-layer.

2.2. Multilayer Undercut Masters

This work was motivated by the wing surface of the Morpho butterfly, which is a periodic, multilayered ridge on the scales of the butterfly's wings [1,38]. To achieve multilayer undercut structures to resemble those found in the Morpho butterflies, the fabrication approach was adopted in three ways:

Firstly, as an etch mask, a lithographically patterned gold layer was used instead of the laser-patterned PTM layer. For lithography either (i) optical photolithography, (ii) electron beam lithography with 350 nm broad features, or (iii) nanoimprint lithography with 300 nm features were employed. For photolithography, AZ5214E (Microchemicals, Ulm, Germany) was used in image reversal mode to generate 1.6 µm high resist patterns for the subsequent lift-off process. For electron beam lithography the PMMA-resist 950K A4 with a thickness of 220 nm was used, and the exposure of lines with a width of 350 nm was performed with a Raith e-line system at 10 kV and a beam current of 200 pA. With electron beam lithography nanostructuring of larger areas requires inefficiently long exposure time. As an alternative for the nanopatterning of larger areas exceeding 1 cm^2 NIL has proven to be a faster process [39]. For the NIL-made etch template a stamp with a checkerboard-style array of 300×300 nm^2 mesas was used to imprint these square structures into a NIL material layer. For fabrication of a metal nanomesh array a dual layer lift-off nanoimprint process using LOR1A (MicroChem Corp., Westborough, MA, USA) and mr-NIL212FC resist materials (*micro resist technology* GmbH, Berlin, Germany) was used as described in [40].

A gold layer with 200 nm nominal thickness was deposited on the lithographic pattern by sputtering with 25 W power using a VonArdenne (Dresden, Germany) LS320 Sputter deposition system or by thermal evaporation using a Balzer's BAK600 (Oerlikon Balzers Coating AG, Brügg, Switzerland) thermal evaporation system. In both cases, a sub-10 nm Cr layer was deposited as an adhesion layer on the substrate. By lift-off, the metal hard mask for the further NIL master fabrication was obtained.

Secondly, instead of using a single Si layer, a stack of alternating Si-SiO_2 layers was used, as sketched out in Figure 4b. The sputter deposited layer stack was comprised of 6 pairs of a 50 nm thick amorphous Si layer on top of an 85 nm thick SiO_2 layer. For a multilayer structure, the individual Si-layers have to be separated by interspaces that can also be etched (to continue etching of the subsequently lower Si-layer). While the amorphous Si was the layer to be underetched sideways, the interspace layer should display a significantly lower etch rate sideways.

Therefore, thirdly, instead of the pure wet chemical etching a reactive ion etching (RIE) process was used. Dry etching was performed in an Oxford Instruments (Oxford Instruments Plasma Technology, Yatton, UK) Plasmalab 100 system using the process gases SF_6, O_2, and Ar. The process pressure was varied between 5 and 60 mTorr to obtain optimal etching results. Si and SiO_2 can both be dry etched by reactive ion etching (RIE) with a fluorine species, but the etch rate of SiO_2 is significantly smaller. The PTM layer (as used in the previous section) is not a suitable mask layer for RIE, as it can only be wet etched after phase transition by blue light exposure.

The RIE process was optimized as a sequence of two etch steps: The first etch step was designed to etch anisotropically through the Si/SiO_2-stack. Hence, this first step was performed at 40 W RF power with a low chamber pressure (5 mTorr), for a higher bias voltage and longer mean free path length. Additionally, to quench the etch selectivity of Si over SiO_2 during the anisotropic etch phase, also oxygen was added to the fluorine etchant (5 sccm SF_6, 24 sccm O_2, 1 sccm Ar). The oxidation of the surface of the Si-layer resulted in quenching the otherwise higher etch rate of Si in contrast to SiO_2 and avoided a side etch into the Si during the anisotropic etch phase. Process step duration for anisotropic etching was in the range of 20 min to 60 min depending on the number of layers. In the second etch step (40 W RF), the underetching was performed by using a strongly isotropic etch process. In this second step, more fluorine and less oxygen was added (20 sccm SF_6, 1 sccm O_2, 9 sccm Ar), and the etch process with the F-species was performed at higher chamber pressure (30 mTorr). Tuned in such a way, this RIE process achieved the under-etching into the Si-layers within 30 s. The resulting master stamp structures were inspected by scanning electron microscopy (SEM) (LEO 1530 VP, ZEISS Microscopy Jena, Germany). To recognize and to measure the undercut either SEM imaging with a tilted view angle or of a cross-section through the undercut structure was performed. Cross-sections were prepared by ion milling with a Zeiss Neon 40 EsB focused ion beam system (ZEISS Microscopy, Jena, Germany).

3. Results

The single-layer T-shaped masters demonstrate the feasibility to fabricate undercut structures also on a large scale. The PTM hard mask [41] was manufactured by Blu-Ray mastering, which is an industrial production technique used for optical memory devices of 120 mm diameter. For these Blu-Ray discs also industrial fabrication equipment exists and fabrication of large quantities has been proven. Using the same industrial fabrication environment for NIL masters opens a quickly accessible supply route for NIL masters at low costs. Figure 1 shows the successful realization of a NIL master with a single layer undercut structure.

As test structures, four squares of 25 × 25 mm were patterned (see Figure 1c). The Blu-Ray-technology-based patterning yielded sub-µm structures that serve as an etch mask and subsequently as the suspended lamellae. Etching of the exposed PTM material revealed a line edge roughness in the range of 20 nm. This is assumedly due to the spacing between the writing tracks, which leads to a curved front of phase transition. For exposure, the orthogonal pattern of our master had to be translated into exposure intervals on the spiral-shaped tracks that the laser was writing on the rotating substrate.

The patterned PTM layer was used as a hard mask to etch the aSi-layer in a wet isotropic KOH etch. As shown in the FIB-cross-section in Figure 1a the sideways underetch was in the range of 150 nm. These single-layer undercut structures were successfully produced over an area of several cm^2 (see Figure 2a). During etching of the Si-layer also the line edge roughness from the PTM layer is transferred into the Si-layer. The FIB-cross section of the underetched Si layer also shows that the distance of underetch is often not identical on the left and the right side (Figure 1a). With wet chemical etching, the diffusion in the underetched cavities is critical, and concentration gradient may lead to inhomogeneous etch rates on opposing sides of the same structure. Despite this restriction, the overall structure fidelity was good over large areas (Figure 1b). Preliminary experiments also demonstrated the feasibility to replicate the inverse structure by NIL. During demolding, the master remained widely intact, which indicates the excellent adhesion of the PTM-layer on the Si-layer (Figure 2a). Damage to the master structures was only evident on few positions throughout the substrate.

For the single layer undercut structure the PTM layer was both (a) laser-patterned hard mask on top and (b) etch stop layer below the Si-layer, which was to be underetched by KOH. Alternatively, also a RIE-process was investigated to avoid the complications of wet chemical etching. Figure 3a shows the same structure as Figure 1 only that the underetching was performed by dry etching. It can be seen in Figure 3a that also the PTM etch stop layer for KOH has been etched, so that the etching continued into the Si substrate. For the T-structure, it is evident, that the top PTM-structure was diminished from an initial 50 nm thickness to a thickness below 20 nm. The result was taken as stimulus to also use a quadruple stack of PTM-layer on top of a Si-layer. Etching such a multi-stack allowed to fabricate a dual layer undercut master and to demonstrate the feasibility to produce more complex undercut structures. However, the fluorine-based RIE process also etches the PTM hard mask on top (Figure 3b).

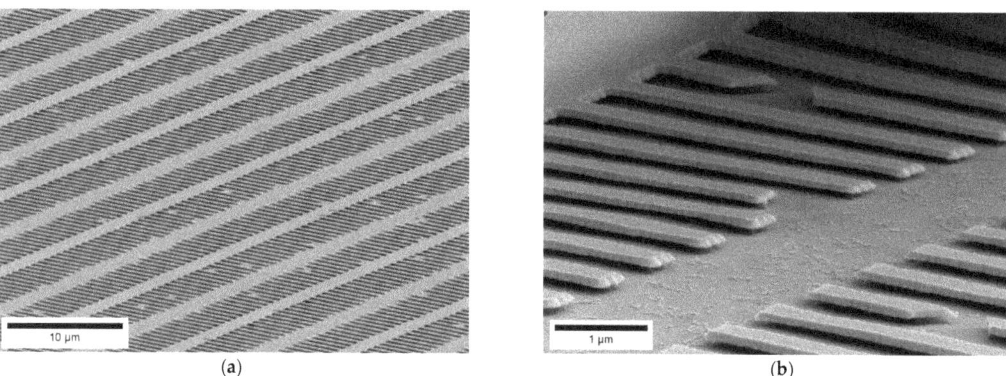

Figure 2. Defects on master after demolding from the NIL-substrate. (a) Overview image and (b) close-up of defect on master.

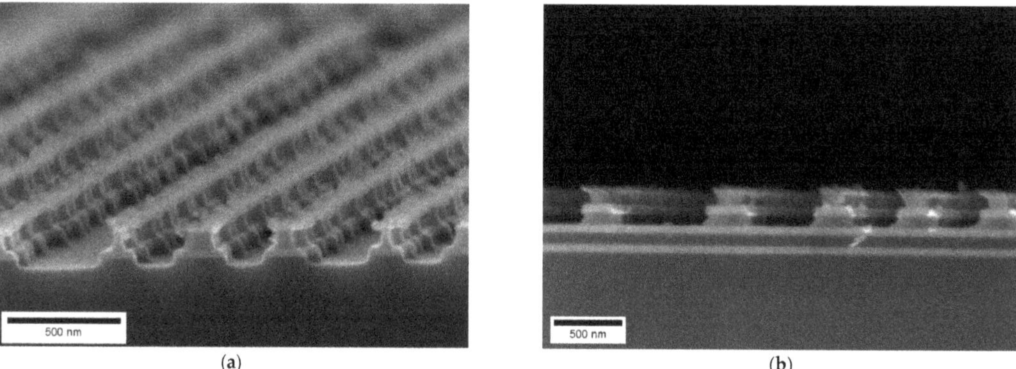

Figure 3. A PTM-Si-stack is etched by a F-based RIE process. (a) single layer PTM/Si-stack and (b) quadruple layer PTM/Si-stack. The cross-sections were fabricated by cleaving the wafer. Images were recorded by SEM on a vertically mounted master chip.

As PTM is both, the pattern defining hard mask and the interlayer between aSi-layers, the RIE process equally removes the PTM interlayer and the pattern-defining hard mask. We conclude that with the given material stack and the single-step RIE process no complex multilevel underetched structures are feasible.

For this reason, we progressed to a different material system—aSi and SiO_2—for the multilayer undercut masters. Due to a size incompatibility of the PTM sputter systems with the SiO_2 sputter system, patterning by Blu-Ray mastering had to be replaced by conventional lithography. As hard mask material, a gold layer was used. The hard mask for the sample shown in Figure 4a was a gold layer prepared by optical lithography and lift off. The multi-stack of Si and SiO_2 is schematically illustrated in Figure 4b. The etching process was optimized by tuning the pressure and the mixture ratio of the etch gases SF_6, O_2, and Ar. By changing the gas composition, the pressure and the inductive power, a change from anisotropic to isotropic etching could be achieved. In the first RIE etch step the layer stack was anisotropically etched, while in a second etching step the Si-layers were isotropically etched sideways, while the SiO_2 interspace layers were not etched significantly. This resulted in the multilevel underetched structure shown in Figure 4. The hard mask for the sample shown in Figure 4a was a gold layer prepared by optical lithography and lift-off. Even after etching the gold layer is still present. This gives rise to the expectation, that even thicker layers or more layers could be processed by the same process. The high anisotropy of the first RIE step can be recognized by the vertical escape line of the corners of the SiO_2 layers in Figure 4a. The underetch into the Si-layer was in the range of 300 nm in top Si-layers down to 150 nm in bottom Si-layers. This gradually decreasing underetch width is analogous to the wing structure on Morpho butterfly. The wing structures are not in the μm range but the sub-μm range. Hence, smaller patterns with feature sizes in the 500 nm to 100 nm range were fabricated.

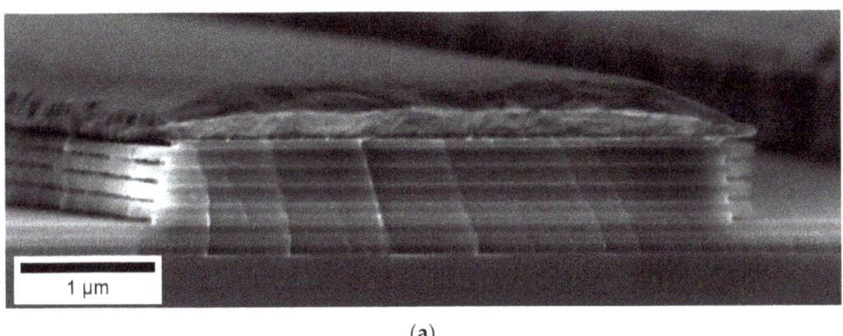

(a)

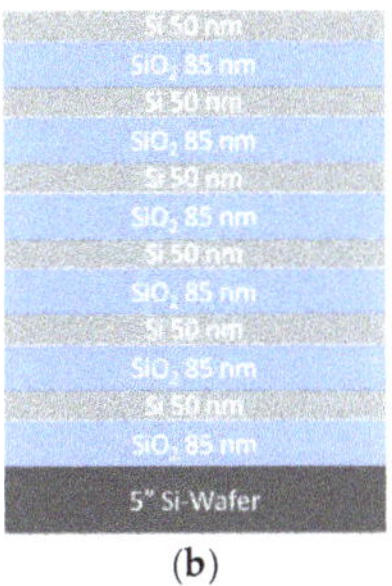

(b)

Figure 4. (**a**) A multilevel underetched structure with a gold mask defined by optical lithography. The lithographic structure is 5 μm wide. SEM image on the cross-section of the layer stack after the dual-stage etching (first etch step stopped after Si layer #5). (**b**) Schematic sequence of layers (without the gold hard mask).

To achieve these sub-μm structures we used either electron beam lithography or nanoimprint lithography to prepare the hard mask pattern. An example obtained using e-beam lithography is shown in Figure 5. The sideways underetch distance of Si was in the range of 120 to 200 nm. The remaining ridge of Si was already thin and reached the limit of mechanical stability. By shortening the duration of the second RIE etch step, also less undercut can be achieved. For a densely packed pattern of lines and spaces (Figure 5a) the anisotropic etch process did not yield a perfectly vertical envelope of the SiO_2 lamellae. Increasing the spacing between lines (Figure 5b) yielded vertically aligned corners of the SiO_2 lamellae. Also, here the amount of the sideways underetch into the Si layers decreased from top (more underetch) to bottom (less underetch). While the fabrication of multilevel undercut structures was demonstrated to be feasible also in the nm-range—a range that is relevant for biomimetic structures—the method of electron beam lithography is not suitable for patterning on large areas.

In a final experiment, we developed a process for larger area patterning of multilevel undercut structures. The industrial NIL replication using NIL-steppers and NIL roll-to-roll systems is excellently suited to fabricate 3D-structures on large areas. In this approach, two-dimensional nanoimprint lithography was used to prepare the master for three-dimensional nanoimprint lithography. A NIL stamp with 300 nm features was used to fabricate a gold hard mask analogous to the previous experiments (see Figures 4 and 5). By a two-step RIE procedure this two-dimensional checkerboard pattern of the gold hardmask was transferred into a three-dimensional structure of the multilevel undercut master. Concluding, a three-dimensional NIL master was fabricated using a two-dimensional NIL-patterned hard mask by etching and is shown in Figure 6.

The FIB-milled cross-section reveals four layers of undercut structures. Even with these 300 nm structures, a clear undercut could be achieved. For the top Si layer, the width of the remaining Si is below 50 nm, which is already at the limit of the mechanical stability of the master. Similar to the Morpho butterfly wing structure, the structures get wider from top to bottom. With 300 nm features the resolution required for biomimicry of the blue butterfly wing of Morpho menelaus could be achieved. The repeated replication of NIL masters with multilayer undercut T-structures has already been demonstrated in [42] and marks an important step towards the reproducible fabrication of biomimetic nanostructures. With the availability of a production procedure for multilevel undercut master structures, the first step towards the technological use of NIL for structural colors has been made.

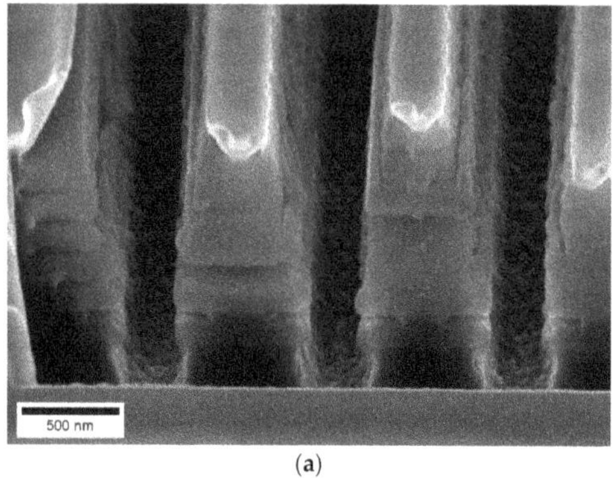
(a)

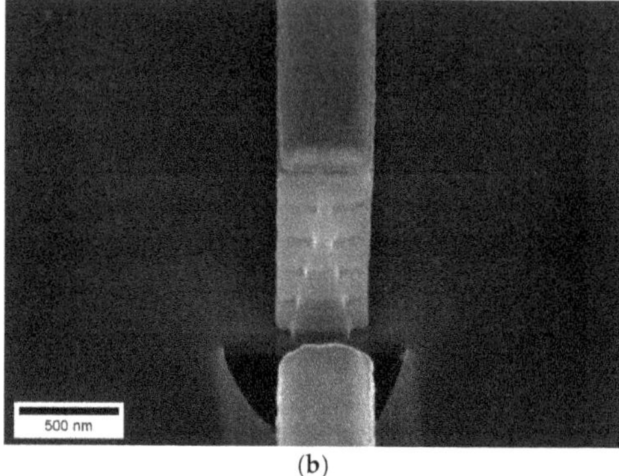
(b)

Figure 5. A multilevel underetched structure with a gold hard mask defined by electron beam lithography. (**a**) SEM image on the cross-section of the layer stack after etching. (**a**) Densely packed lines with little spacing. Lithographic structures are 400 nm (middle-left) and 350 nm (middle-right) wide. (**b**) Line with more than 1 μm inter-line-distance. The lithographic structure is 350 nm wide.

(a)

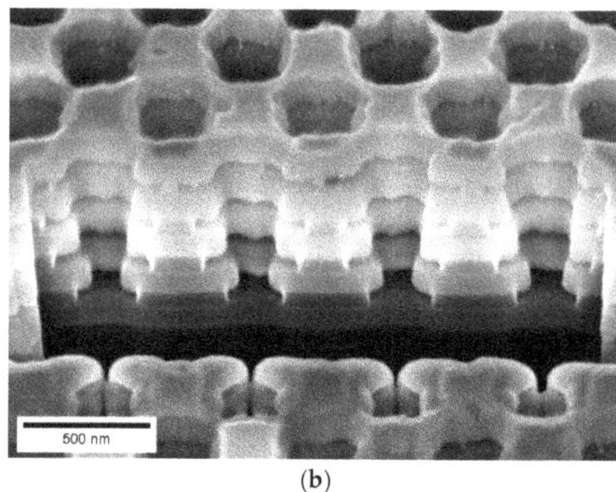
(b)

Figure 6. A multilevel underetched structure with a gold hard mask was defined by a NIL imprinted stamp. A 2D checkerboard structure consisting of 300×300 nm^2 squares was transferred into a metal layer acting as etch hardmask. The checkerboard pattern was transferred into the material by RIE etching. (**a**) Overview; (**b**) close-up.

4. Conclusions

Nanoimprint lithography is an exceptional patterning technique, as it can directly replicate three-dimensional structures. The availability of a three-dimensional master is a bottleneck of the NIL replication. In this work, we have presented fabrication technologies for undercut three-dimensional masters. With the structural design of our masters, we were inspired by the structural color of Morpho butterfly wings.

We have successfully demonstrated different ways to fabricate undercut nanoscale features on large areas. In a first step, a two-dimensional hard mask was prepared on a multilayer stack. In a subsequent two-stage RIE-process the undercut features were fabricated. The use of PTM hard masks allows using the industrially established method of laser writing as used for Blu-Ray mastering. Alternatively, also optical lithography, electron beam lithography and even nanoimprint lithography itself have been successfully demonstrated as alternative approaches to fabricate hard mask layers. With the industrially preferred RIE process, the problem-prone wet etching can be avoided. With the two-stage RIE process an industrially viable route to produce multilevel undercut structures was demonstrated. With an optimized process we could successfully show the fabrication of a four-layer undercut structure with 300×300 nm features.

In future generations these approaches—PTM hard mask fabrication by Blu-Ray mastering and multilayer undercut etching by dual-stage RIE—may be combined. This will yield an industrially relevant fabrication approach for complex three-dimensional masters. The multilevel undercut master structures produced in this work are an essential component to further develop an imprinting and demolding procedure for complex undercut structures. NIL steppers and NIL roll-to-roll systems may transfer this approach to the industrial fabrication of structural colors on large areas.

Author Contributions: Conceptualization, P.T., A.P., H.D.W. and M.M.; Investigation, P.T., A.P., H.D.W., P.S., M.S., A.T., M.M.S. and M.H.; Writing—original draft preparation, H.D.W.; Writing—review and editing, H.D.W. and P.T.; Supervision, H.D.W.; Project administration, M.M. and H.D.W.; Funding acquisition, M.M. and H.D.W. All authors have read and agreed to the published version of the manuscript.

Funding: This research was funded by the Austrian Research Promotion Agency (FFG) under FFG-grant 843639 (project: "rollerNIL"). Open Access Funding by TU Wien.

Data Availability Statement: The data presented in this study are available on request from the corresponding author after agreeing to non-disclosure. The raw data are not publicly available due to non-disclosure agreements within the grant consortium.

Conflicts of Interest: The authors declare no conflict of interest.

References

1. Vukusic, P.; Sambles, J.R.; Lawrence, C.R.; Wootton, R.J. Quantified Interference and Diffraction in Single Morpho Butterfly Scales. *Proc. R. Soc. B Biol. Sci.* **1999**, *266*, 1403–1411. [CrossRef]
2. Vukusic, P.; Sambles, J.R. Photonic Structures in Biology. *Nature* **2003**, *424*, 852–855. [CrossRef] [PubMed]
3. Fernandez, A.; Francone, A.; Thamdrup, L.H.; Johanson, A.; Bilenberg, B.; Nielsen, T.; Guttmann, M.; Torres, C.M.S.; Kehagias, N. Design of Hierarchical Surfaces for Tuning Wetting Characteristics. *Acs Appl. Mater. Interfaces* **2017**, *9*, 7701–7709. [CrossRef] [PubMed]
4. Kirchner, R.; Guzenko, V.A.; Rohn, M.; Sonntag, E.; Mühlberger, M.; Bergmair, I.; Schift, H. Bio-Inspired 3D Funnel Structures Made by Grayscale Electron-Beam Patterning and Selective Topography Equilibration. *Microelectron. Eng.* **2015**, *141*, 107–111. [CrossRef]
5. Liddle, J.A.; Gallatin, G.M. Nanomanufacturing: A Perspective. *ACS Nano* **2016**, *10*, 2995–3014. [CrossRef]
6. Jiang, T.; Guo, Z.G.; Liu, W.M. Biomimetic Superoleophobic Surfaces: Focusing on Their Fabrication and Applications. *J. Mater. Chem. A* **2015**, *3*, 1811–1827. [CrossRef]
7. Luong-Van, E.; Rodriguez, I.; Low, H.Y.; Elmouelhi, N.; Lowenhaupt, B.; Natarajan, S.; Lim, C.T.; Prajapati, R.; Vyakarnam, M.; Cooper, K. Review: Micro-and Nanostructured Surface Engineering for Biomedical Applications. *J. Mater. Res.* **2013**, *28*, 165–174. [CrossRef]
8. Khan, S.; Lorenzelli, L.; Dahiya, R.S. Technologies for Printing Sensors and Electronics over Large Flexible Substrates: A Review. *IEEE Sens. J.* **2015**, *15*, 3164–3185. [CrossRef]
9. Busa, C.; Rickard, J.J.S.; Chun, E.; Chong, Y.; Navaratnam, V.; Oppenheimer, P.G. Tunable Superapolar Lotus-to-Rose Hierarchical Nanosurfaces via Vertical Carbon Nanotubes Driven Electrohydrodynamic Lithography. *Nanoscale* **2017**, *9*, 1625–1636. [CrossRef] [PubMed]
10. Schift, H. Nanoimprint Lithography: An Old Story in Modern Times? A Review. *J. Vac. Sci. Technol. B Microelectron. Nanometer Struct.* **2008**, *26*, 458–480. [CrossRef]
11. Möllenbeck, S.; Bogdanski, N.; Wissen, M.; Scheer, H.C.; Zajadacz, J.; Zimmer, K. Investigation of the Separation of 3D-Structures with Undercuts. *Microelectron. Eng.* **2007**, *84*, 1007–1010. [CrossRef]
12. Goh, W.H.; Hashimoto, M. Fabrication of 3D Microfluidic Channels and In-Channel Features Using 3D Printed, Water-Soluble Sacrificial Mold. *Macromol. Mater. Eng.* **2018**, *303*. [CrossRef]
13. Matsumoto, H.; Okabe, T.; Taniguchi, J. Microchannel Fabrication via Ultraviolet-Nanoimprint Lithography and Electron-Beam Lithography Using an Ultraviolet-Curable Positive-Tone Electron-Beam Resist. *Microelectron. Eng.* **2020**, *226*, 111278. [CrossRef]

14. An, J.H.; Choi, J.S.; Kang, S.M. Reliable replication molding process for robust mushroom-shaped microstructures. *J. Korean Soc. Precis. Eng.* **2020**, *37*, 855–860. [CrossRef]

15. Finn, A.; Hensel, R.; Hagemann, F.; Kirchner, R.; Jahn, A.; Fischer, W.-J. Geometrical Properties of Multilayer Nano-Imprint-Lithography Molds for Optical Applications. *Microelectron. Eng.* **2012**, *98*, 284–287. [CrossRef]

16. Kirchner, R.; Chidambaram, N.; Schift, H. Benchmarking Surface Selective Vacuum Ultraviolet and Thermal Postprocessing of Thermoplastics for Ultrasmooth 3-D-Printed Micro-Optics. *Opt. Eng.* **2018**, *57*. [CrossRef]

17. Zimmer, K.; Zajadacz, J.; Fechner, R.; Dhima, K.; Scheer, H.-C. Fabrication of Optimized 3D Microstructures with Undercuts in Fused Silica for Replication. *Microelectron. Eng.* **2012**, *98*, 163–166. [CrossRef]

18. Mayer, A.; Rond, J.; Staabs, J.; Leifels, M.; Zajadacz, J.; Ehrhardt, M.; Lorenz, P.; Sunagawa, H.; Hirai, Y.; Zimmer, K.; et al. Multiple Replication of Hierarchical Structures from Polymer Masters with Anisotropy. *J. Vac. Sci. Technol. B Nanotechnol. Microelectron.* **2019**, *37*. [CrossRef]

19. Steinberg, C.; Rumler, M.; Runkel, M.; Papenheim, M.; Wang, S.; Mayer, A.; Becker, M.; Rommel, M.; Scheer, H.-C. Complex 3D Structures via Double Imprint of Hybrid Structures and Sacrificial Mould Techniques. *Microelectron. Eng.* **2017**, *176*, 22–27. [CrossRef]

20. Okabe, T.; Maebashi, H.; Taniguchi, J. Ultra-Violet Nanoimprint Lithography-Compatible Positive-Tone Electron Beam Resist for 3D Hybrid Nanostructures. *Microelectron. Eng.* **2019**, *213*, 6–12. [CrossRef]

21. Alameda, M.T.; Osorio, M.R.; Hernández, J.J.; Rodríguez, I. Multilevel Hierarchical Topographies by Combined Photolithography and Nanoimprinting Processes to Create Surfaces with Controlled Wetting. *Acs Appl. Nano Mater.* **2019**, *2*, 4727–4733. [CrossRef]

22. Moharana, A.R.; Außerhuber, H.M.; Mitteramskogler, T.; Haslinger, M.J.; Mühlberger, M.M. Multilayer Nanoimprinting to Create Hierarchical Stamp Masters for Nanoimprinting of Optical Micro-and Nanostructures. *Coatings* **2020**, *10*, 301. [CrossRef]

23. Deng, J.; Zhou, H.; Dong, J.; Cohen, P. Three-Dimensional Nanomolds Fabrication for Nanoimprint Lithography. In *Procedia Manufacturing*; Fratini, L., Wang, L., Ragai, I., Eds.; Elsevier, B.V.: Amsterdam, The Netherlands, 2019; Volume 34, pp. 228–232.

24. Ofir, Y.; Moran, I.W.; Subramani, C.; Carter, K.R.; Rotello, V.M. Nanoimprint Lithography for Functional Three-Dimensional Patterns. *Adv. Mater.* **2010**, *22*, 3608–3614. [CrossRef] [PubMed]

25. Mühlberger, M.; Rohn, M.; Danzberger, J.; Sonntag, E.; Rank, A.; Schumm, L.; Kirchner, R.; Forsich, C.; Gorb, S.; Einwögerer, B.; et al. UV-NIL Fabricated Bio-Inspired Inlays for Injection Molding to Influence the Friction Behavior of Ceramic Surfaces. *Microelectron. Eng.* **2015**, *141*, 140–144. [CrossRef]

26. Waid, S.; Wanzenboeck, H.D.; Muehlberger, M.; Bertagnolli, E. Optimization of 3D Patterning by Ga Implantation and Reactive Ion Etching (RIE) for Nanoimprint Lithography (NIL) Stamp Fabrication. *Microelectron. Eng.* **2012**, *97*, 105–108. [CrossRef]

27. Waid, S.; Wanzenboeck, H.D.; Muehlberger, M.; Gavagnin, M.; Bertagnolli, E. Generation of 3D Nanopatterns with Smooth Surfaces. *Nanotechnology* **2014**, *25*. [CrossRef]

28. Schleunitz, A.; Schift, H. Fabrication of 3D Nanoimprint Stamps with Continuous Reliefs Using Dose-Modulated Electron Beam Lithography and Thermal Reflow. *J. Micromech. Microeng.* **2010**, *20*. [CrossRef]

29. Schrittwieser, S.; Haslinger, M.J.; Mitteramskogler, T.; Mühlberger, M.; Shoshi, A.; Brückl, H.; Bauch, M.; Dimopoulos, T.; Schmid, B.; Schotter, J. Multifunctional Nanostructures and Nanopocket Particles Fabricated by Nanoimprint Lithography. *Nanomaterials* **2019**, *9*, 1790. [CrossRef]

30. Burgsteiner, M.; Müller, F.; Lucyshyn, T.; Kukla, C.; Holzer, C. Influence of the Process Parameters on the Replication of Microstructured Freeform Surfaces. In Proceedings of the AIP Conference Proceedings, Rhodes, Greece, 22–28 September 2014; American Institute of Physics Inc.: College Park, MD, USA, 2014; Volume 1593, pp. 183–188.

31. Schneider, N.; Zeiger, C.; Kolew, A.; Schneider, M.; Leuthold, J.; Hölscher, H.; Worgull, M. Nanothermoforming of Hierarchical Optical Components Utilizing Shape Memory Polymers as Active Molds. *Opt. Mater. Express* **2014**, *4*, 1895–1902. [CrossRef]

32. Meier, T.; Bur, J.; Reinhard, M.; Schneider, M.; Kolew, A.; Worgull, M.; Hölscher, H. Programmable and Self-Demolding Microstructured Molds Fabricated from Shape-Memory Polymers. *J. Micromech. Microeng.* **2015**, *25*. [CrossRef]

33. Yun, J.-G.; Lee, J.D.; Park, B.-G. 3D NAND Flash Memory with Laterally-Recessed Channel (LRC) and Connection Gate Architecture. *Solid-State Electron.* **2011**, *55*, 37–43. [CrossRef]

34. Kim, Y.; Cho, S.; Lee, G.S.; Park, I.H.; Lee, J.D.; Shin, H.; Park, B.G. 3-Dimensional Terraced NAND (3D TNAND) Flash Memory-Stacked Version of Folded NAND Array. *Ieice Trans. Electron.* **2009**, *E92-C*, 653–658. [CrossRef]

35. Kouchiyama, A.; Aratani, K.; Takemoto, Y.; Nakao, T.; Kai, S.; Osato, K.; Nakagawa, K. High Resolution Blue Laser Mastering with Inorganic Photoresist. In Proceedings of the International Symposium on Optical Memory and Optical Data Storage Topical Meeting, Waikoloa, HI, USA, 7–11 July 2002; pp. 27–29.

36. Raoux, S.; Rettner, C.T.; Jordan-Sweet, J.L.; Kellock, A.J.; Topuria, T.; Rice, P.M.; Miller, D.C. Direct Observation of Amorphous to Crystalline Phase Transitions in Nanoparticle Arrays of Phase Change Materials. *J. Appl. Phys.* **2007**, *102*. [CrossRef]

37. Meinders, E.R.; Rastogi, R.; Van Der Veer, M.; Peeters, P.; El Majdoubi, H.; Bulle, H.; Millet, A.; Bruls, D. Phase-Transition Mastering of High-Density Optical Media. *Jpn. J. Appl. Phys. Part 1 Regul. Pap. Short Notes Rev. Pap.* **2007**, *46*, 3987–3992. [CrossRef]

38. Giraldo, M.A.; Yoshioka, S.; Liu, C.; Stavenga, D.G. Coloration Mechanisms and Phylogeny of Morpho Butterflies. *J. Exp. Biol.* **2016**, *219*, 3936–3944. [CrossRef]

39. Bergmair, I.; Dastmalchi, B.; Bergmair, M.; Saeed, A.; Hilber, W.; Hesser, G.; Helgert, C.; Pshenay-Severin, E.; Pertsch, T.; Kley, E.B.; et al. Single and Multilayer Metamaterials Fabricated by Nanoimprint Lithography. *Nanotechnology* **2011**, *22*. [CrossRef] [PubMed]
40. Haslinger, M.J.; Mitteramskogler, T.; Kopp, S.; Leichtfried, H.; Messerschmidt, M.; Thesen, M.W.; Mühlberger, M. Development of a Soft UV-NIL Step&repeat and Lift-off Process Chain for High Speed Metal Nanomesh Fabrication. *Nanotechnology* **2020**, *31*. [CrossRef]
41. Wuttig, M.; Yamada, N. Phase-Change Materials for Rewriteable Data Storage. *Nat. Mater.* **2007**, *6*, 824–832. [CrossRef] [PubMed]
42. Mühlberger, M.; Stephan, S.R.; Nees, D.; Moharana, A.; Belegratis, M.R.; Taus, P.; Kopp, S.; Wanzenboeck, H.D.; Prinz, A.; Fechtig, D. Nanoimprint Replication of Biomimetic, Multilevel Undercut Nanostructures. *Nanomaterials* **2021**. accepted.

 *nanomaterials* 

Article

Nanoimprint Replication of Biomimetic, Multilevel Undercut Nanostructures

Michael Muehlberger [1,*], Stephan Ruttloff [2], Dieter Nees [2], Amiya Moharana [1,2], Maria R. Belegratis [2], Philipp Taus [3], Sonja Kopp [1], Heinz D. Wanzenboeck [3], Adrian Prinz [4] and Daniel Fechtig [1]

1 Profactor GmbH, Im Stadtgut D1, 4407 Steyr, Austria; Amiya.Moharana@joanneum.at (A.M.); Sonja.Kopp@profactor.at (S.K.); Daniel.Fechtig@profactor.at (D.F.)
2 JOANNEUM RESEARCH Forschungsgesellschaft mbH, 8160 Weiz, Austria; Stephan.Ruttloff@joanneum.at (S.R.); Dieter.Nees@joanneum.at (D.N.); Maria.Belegratis@joanneum.at (M.R.B.)
3 Institute for Solid State Electronics, TU Wien, 1040 Vienna, Austria; Philipp.Taus@tuwien.ac.at (P.T.); Heinz.Wanzenboeck@tuwien.ac.at (H.D.W.)
4 Stratec Consumables GmbH, 5081 Anif, Austria; A.Prinz@stratec.com
* Correspondence: Michael.Muehlberger@profactor.at

Abstract: The nanoimprint replication of biomimetic nanostructures can be interesting for a wide range of applications. We demonstrate the process chain for Morpho-blue-inspired nanostructures, which are especially challenging for the nanoimprint process, since they consist of multilayer undercut structures, which typically cannot be replicated using nanoimprint lithography. To achieve this, we used a specially made, proprietary imprint material to firstly allow successful stamp fabrication from an undercut master structure, and secondly to enable UV-based nanoimprinting using the same material. Nanoimprinting was performed on polymer substrates with stamps on polymer backplanes to be compatible with roller-based imprinting processes. We started with single layer undercut structures to finally show that it is possible to successfully replicate a multilayer undercut stamp from a multilayer undercut master and use this stamp to obtain multilayer undercut nanoimprinted samples.

Keywords: nanoimprint lithography; UV-NIL; biomimetics; morpho butterfly

Citation: Muehlberger, M.; Ruttloff, S.; Nees, D.; Moharana, A.; Belegratis, M.R.; Taus, P.; Kopp, S.; Wanzenboeck, H.D.; Prinz, A.; Fechtig, D. Nanoimprint Replication of Biomimetic, Multilevel Undercut Nanostructures. *Nanomaterials* **2021**, *11*, 1051. https://doi.org/10.3390/nano11041051

Academic Editor: Konstantins Jefimovs

Received: 24 February 2021
Accepted: 9 April 2021
Published: 20 April 2021

Publisher's Note: MDPI stays neutral with regard to jurisdictional claims in published maps and institutional affiliations.

1. Introduction

In the course of evolution, nature has developed a broad range of nanostructured surfaces with many different purposes: to optimize friction behavior (e.g., [1–3]) or the behavior of liquids (e.g., [4,5]), to achieve antimicrobial (e.g., [6,7]) or antireflective effects (e.g., [8–10]) and to achieve decorative effects (e.g., [11–14]), just to name a few. Many of these nanostructures are also of high interest for technical applications and have been investigated accordingly (e.g., [1,15–17]).

Nanoimprinting [18–20] is a replication technology, capable of replicating nanoscale (e.g., [21,22]) as well as microscale features [23] on large areas [24–26].

Due to the fact that nanoimprinting can—in a single process step—replicate complex multilevel structures [1,24,27,28], it is ideally suited for the replication of such biomimetic or bio-inspired nanostructures. A big challenge for nanoimprinting, however, is the replication of undercut features, since it is usually not possible to remove the stamp from such features, especially if they are of nanoscale. It has, however, been shown in the literature that this can be achieved for larger features in the several μm range [29,30].

The nanostructures we are interested in are responsible for the metallic blue color of the Morpho butterfly. These structures have attracted considerable interest and their functioning has been researched in detail [31,32]. Based on the understanding of the optical properties of these nanostructures, it has been possible to mimic the optical effect using nanostructured surfaces and multilayer stacks of deposited materials [31]. We wanted to investigate if it might be possible to avoid such complicated and expensive partly vacuum-based processes and replace them with a single nanoimprint process. The aim of the work presented in this paper was, however, not to replicate the full optical effect, but to demonstrate the basic possibility. The mastering process would have been significantly more complicated if we would have included the necessary randomness to its full extent as it would have been necessary to replicate the full optical effect. The capabilities of the nanoimprint process can, however, also be shown with a reduced master design.

2. Materials and Methods

2.1. Master Design and Fabrication

Figure 1 schematically shows cross-sections of the structures that we dealt with in this work. The very left structure represents the structure as it is found in nature and is drawn after SEM images as they are found in the literature (e.g., [12,33]). For our purpose, we repeated the individual structures periodically along the x-direction. The structures extend along the y direction, as can also be seen later in Figure 6 (left image).

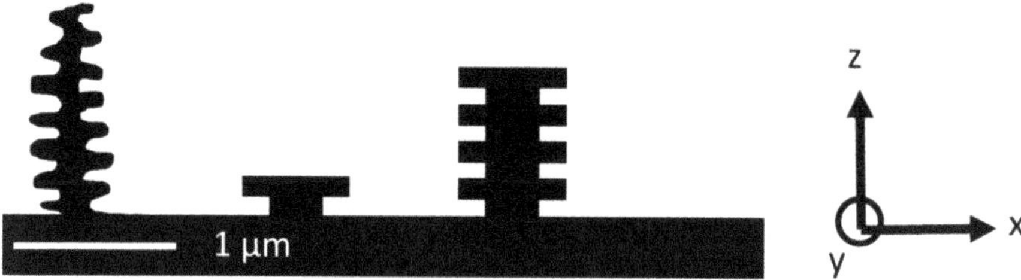

Figure 1. Left: sketch of the cross-section of the nanostructures responsible for the metallic-blue color of the Morpho butterfly. The sketch was created after images found, e.g., in [12,33]. Center and right: model structures used in this work. We started with a single layer "T-structure" and continued to work with the more complex "tree-structure".

As far as we understand from the literature [34], to obtain the full optical effect of the Morpho butterfly in our model structures, several aspects are missing, most notably randomness in the lateral placement of the individual "trees" as well as randomness in the height of the "trees". Furthermore, the branches of the trees are not positioned at the same height on the left and on the right side of the trunk in the real Morpho butterfly structure. Taking these facts into account for our master design would have significantly increased the challenges in master fabrication. At the same time, the nanoimprint challenge would have been basically the same. Therefore, since the goal was to investigate if nanoimprinting can be used to replicate such a structure, we focused on the model structures as sketched in Figure 1 in the center and on the right and did not include the randomness.

Details on the masters and the mastering process are described in [35]; here, only the basic facts should be mentioned. All masters were fabricated on silicon wafers, first using a multilayer deposition followed by lithography and anisotropic etching to etch through the multilayer stack. Additionally, one of the layer materials was selectively etched to create the undercut. The layer heights of the multilayer stack were 85 nm and 50 nm. The lateral width of the single T-structure was around 400 nm after processing, while tree structures with different lateral dimensions were fabricated. Undercuts in the range of 100 nm were typically achieved.

2.2. Nanoimprint Material

The imprint resin primarily used in this work is an in house development of JOAN-NEUM RESEARCH Forschungsgesellschaft mbH. In the development process, 13 different acrylate resins have been formulated and tested, with the main focus on low viscosity, high elasticity and low surface energy in order to enable the fast and complete filling of the undercut features and easy demolding of the cured structures. Some of these resins had problems during demolding and stuck to the stamp, e.g., leaving residual material in the undercut areas. Others demolded easily but did not replicate the undercut features. Typical defects were collapsed T-structures. Out of the tested ones, the resin named NILcure® JR5 (JOANNEUM RESEARCH, Weiz, Austria) proved to be optimal for high fidelity replication of undercut structures.

This elastic UV curable acrylate resin can be efficiently cured within seconds by UV-A light between 350 nm and 400 nm at a moderate irradiance of 50–100 mW/cm². Additional properties facilitating the UV imprinting of undercut features are low viscosity (<20 m Pas), low glass transition temperature (below room temperature), high crosslinking and low surface energy after curing down to 20 mN/m—depending on the surface energy of the stamp [25]. Furthermore, this imprint material shows high scratch and abrasion resistance as well as chemical and weathering stability.

A key feature of NILcure® JR5 that was used in the course of this project is also that it can be used as a stamp material for NILcure® JR5 itself, meaning that it is self-replicable. The nanoimprint process using this material is described in the next section.

The challenge was to find the right balance between surface energy and mechanical properties, i.e., Young's modulus, tensile strength and elongation at break of the cured imprint resin. The surface energy of the resins was tuned by doping the resins with highly surface active anti-adhesive additives—such as 1H,1H,2H,2H-perfluorooctyl-acrylate—as well as a JR-proprietary related molecule.

For tuning the viscosity of the uncured and the mechanical properties of the cured imprint resins, various formulations containing multifunctional acrylate oligomers with aliphatic poly-urethane and poly-siloxane backbones as well as mono- and bifunctional acrylate monomers were investigated. The acrylate oligomers were chosen according to their mechanical properties, ranging from Young's Modulus = 20 MPa–100 MPa, tensile strength = 5–25 MPa and elongation at break = 10–100%. For formulating the UV-imprint resins, reactive diluents had to be added to the highly viscos acrylate-functionalized oligomers. Mono- as well as bi-functional acrylate monomers were chosen. Whilst the addition of mono-functional acrylate monomers largely preserves low Young's modulus and high elongation at break, the tensile strength is generally lowered. The addition of bifunctional acrylate monomers, on the other hand, decreases the elongation at break and increases the Young's modulus and tensile strength of the cured resins [36,37].

NILcure® JR5 featuring optimal mechanical properties and the JR proprietary fluorinated anti-adhesive additive turned out to enable the highest imprint fidelity for the undercut structures of all investigated resin formulations.

A universal testing machine (Instron 3342 [38]) was used to determine the mechanical parameters of the most successful material NILcure® JR5 (see Figure 2). This machine can be used to perform classical tensile tests (Young's modulus, tensile strength, elongation at break) as well as other important measurements such as friction and adhesion measurements (peel test). For the determination of the Young's modulus and the elongation at break (e.g., [39]), a classical tensile test was performed. The rectangular tensile specimens had dimensions of 40 mm × 10 mm × 150 µm (clamping length, width, thickness). The test speed was 5 mm/min. The following figure shows an exemplary measurement of JR5. The following parameters were obtained: Young's modulus: 246 MPa, elongation at break: 8%. The surface energy of the liquid material is 25.5 mN/m as determined by the pendant drop method [40,41], using a DSA100 from Krüss [42].

Initially, materials such as h-PDMS (hard Polydimethylsiloxane) [43] and MD700 [44] as stamp materials and OrmoComp [45] as an imprint material were used. However, the results were not encouraging in terms of defectivity (see later in Figure 4), which is why our own materials were formulated and tested. The results are discussed below in the results section.

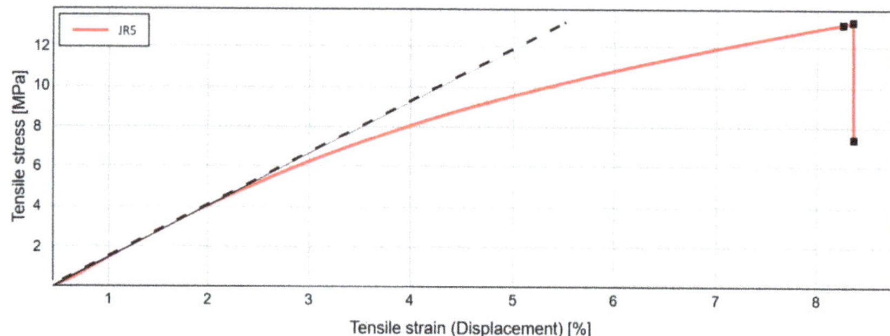

Figure 2. Tensile stress plotted versus tensile strain for the NILcure® JR5 material.

2.3. Nanoimprint Process

The goal of our nanoimprint process was to replicate a complex undercut nanostructure in a single imprinting step. To do so, we fabricated a stamp from a nanoimprint master and used this stamp in a nanoimprint process. For all process steps, we used UV-based nanoimprinting, i.e., UV-curable materials. The sample fabrication sequence is sketched in Figure 3.

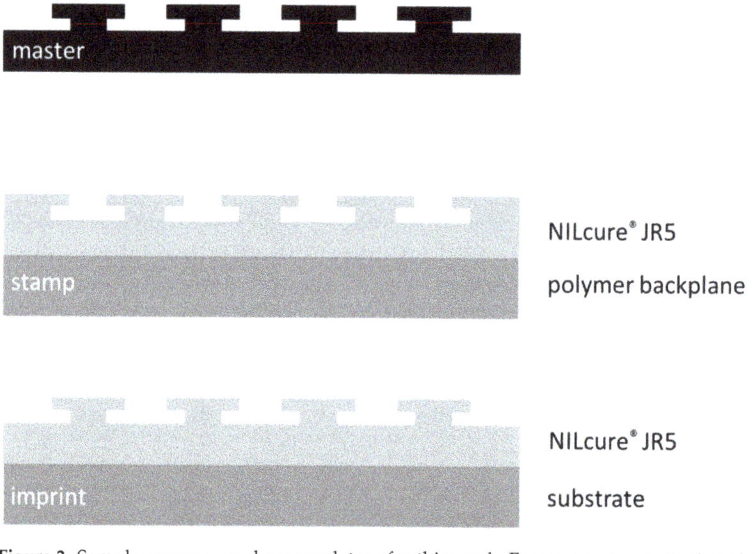

Figure 3. Sample sequence and nomenclature for this work: From a master, a nanoimprint stamp was fabricated (stamp material on backplane), which was then used to create the imprint (imprint material on substrate).

All masters were treated with an anti-sticking layer (BGL-GZ-83 from PROFACTOR GmbH [46,47]) to allow for easy removal of the stamp from the master structure. To fabricate the stamp, NILcure® JR5 resin was manually deposited on the master. A UV-transparent flexible backplane was used and placed gently onto the droplet in a rolling motion alongside the direction of the T-structures (y-direction in Figure 1). The whole stack was UV-exposed with a wavelength of 365 nm and an intensity of about 100 mW/cm^2 for a minimum of 30 s. Both UV-LED (light emitting diode) light sources (self-made at PROFACTOR, 365 nm) and UV fluorescent tubes (supplier: biostep GmbH [48]) can be used. After curing, the stamp was very gently removed from the master, again in the y-direction, to ensure easier demolding and also to prevent the tree branches from being destroyed.

As backplanes, different polymer films have been used. Two hundred micrometer-thick commercial PVC (polyvinyl chloride) films worked well, as well as PET (polyethylene terephthalate) films with a thickness of 125 μm. The PVC was treated with an oxygen plasma before applying an adhesion promoter (HMNP12 from PROFACTOR GmbH [49]). The PET films were already supplied with an adhesion promoter, such as Kemafoil HSPL80 from Coveme [50,51] and Melinex ST506 from DuPont [52]. No backplane was used for the imprints using h-PDMS stamps.

After having obtained the stamps in the way described above, the imprints were made using those stamps in the same way. The imprint material was manually placed on the substrate, the stamp was carefully brought into contact with the coated substrate and, after curing, the stamp was separated from the finished imprint. The substrates that were used were the same polymer foils as for the backplanes for the stamps, but also glass substrates were used (e.g., Superfrost Ultra Plus [53]).

For both processes—the stamp fabrication as well as the actual imprinting—no additional pressure was applied. The imprinting was performed manually and was left to the capillary forces to obtain good contact between the stamp and substrate. The imprinting area was always several mm^2, typically 5×5mm^2.

There are two reasons for using polymer foils in this process chain. Firstly, it was necessary for sample characterization for both stamps and imprints, as described below. Secondly, and this is only valid for the stamp, the flexibility of the stamp allowed for easy separation of the stamp from the imprint in a peeling-like motion. The flexibility and UV-transparency will also allow the use of such stamps as imprinting plates in roller-based nanoimprint processes [25,26], which would pave the way for industrial implementation.

2.4. Sample Characterisation

Sample characterization was performed using atomic force microscopy (AFM) (Bruker Dimension Edge, Bruker Corporation, Billerica, MA, USA) (PROFACTOR) and scanning electron microscopy (TU Wien (Zeiss Neon 40 EsB (Zeiss Microscopy, Jena, Germany) and JOANNEUM RESEARCH (SM-IT 100 (JEOL, Tokyo, Japan)). For the characterization of the master, focused ion beam (FIB) cutting (TU Wien using a Zeiss Neon 40 (Zeiss Microscopy, Jena, Germany) was also used. It proved to be challenging to obtain reliable information on the undercut on the polymer samples from these methods. AFM imaging naturally cannot provide direct evidence of any undercut features, but also indirect evidence gathered by comparing the lateral dimensions of master, stamp and imprint proved to be not conclusive. One reason for this is tip convolution effects. Additionally, standard SEM imaging caused problems since the polymeric nanostructures tend to deform under electron beam bombardment or during FIB cutting.

High-quality cross sections of the nanostructures (in x-direction) on polymer substrates were made by means of cryo-ultramicrotomy (PowerTome XL RMC Products, Boeckeler$^{®}$, Tucson, AZ, USA [54]). The samples were physically hardened by cooling with liquid nitrogen below their glass transition temperature at -60 °C. Nanomaterial imaging and analysis were performed by utilizing the scanning electron microscope RAITH e-line at 3 kV. Cryo-ultramicrotomy at JOANNEUM RESEARCH for sample preparation proved to be the only variant that led to reliable results.

3. Results

3.1. Single Layer Undercut Structures

Using the imprinting and characterization procedures described above, we were able to show the full nanoimprint chain from master to final imprint, for both the single "Ts" as well as the "trees".

In the course of the project, and before the NILcure$^{®}$ JR5 material was used, several other material combinations were tested. Results are shown in Figures 4 and 5. Initially, our approach was to use soft stamps and more rigid imprint materials, such as h-PDMS or MD700 as soft stamp materials and OrmoComp as an imprint material. However, due to the limited tuning possibilities with these commercial materials and the high number of defects (see Figure 4) observed already in the stamps for our special application, we started to test our own materials. In Figure 4, two different material combinations are shown. On the left, the result of an imprint using an h-PDMS stamp and OrmoComp as the imprinting material is shown. As can be seen, the T-shapes are not well defined and the edges of the Ts are rough and irregular, indicating problems during the imprinting step—either non-filling of the undercut or mechanical tear-off of the overhanging part of the nanostructures, or possibly both. A lot of defects can also be seen in the right image of Figure 4, which shows an SEM image of an OrmoComp imprint using an MD700 stamp. Additionally, here the overhanging structures are not well replicated. Here, defects on the bottom can also be observed, which are, however, related to the defective master which was used for these experiments and are not related to the undercut structures.

Figure 4. Left: cross-sectional SEM image of an imprint using an h-PDMS stamp and OrmoComp as imprinting material. Right: SEM image imprint using MD700 as stamp material and OrmoComp as imprint material.

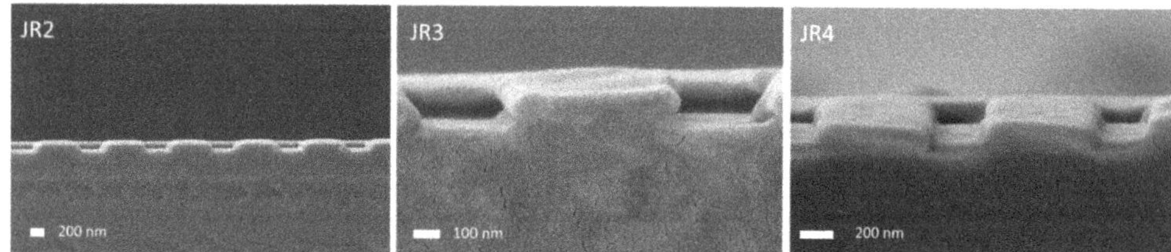

Figure 5. Cross-sectional SEM images of stamps made from different materials. For NILcure® JR2 and JR4 it can be seen that the undercut is not replicated at all, while for JR3 the undercut is only partly replicated.

Switching to the self-made materials described above initially resulted in stamps which did not replicate the undercut, either completely or only partly, as shown in Figure 5, where cross-sectional SEM images of three stamps made from three different experimental materials are shown. It can be seen that the undercut is not replicated (left image and right image) at all, or only partly (center image).

The images in Figures 4 and 5 give a representative overview of the typical defects we observed: either non-replication of the undercut or only partial replication. We attribute the non-replication of the undercut either to non-filling of the undercut part of the structures, which is related to the wetting properties of the material and its viscosity, or to a material which was too soft, so that the nanostructures collapsed. The second type of defect we observe is when the undercut was only partly replicated. This we attribute to an imprint material that was mechanically too brittle to allow for a defect-free separation. The defects described above appear on the whole area with the same density, meaning that the full imprint area cannot be used. This is in contrast to other typical nanoimprint defects induced by dust particles or air bubbles, which are very local. In summary, the mechanical stability of the stamp and the imprint material have to be right as well as the surface energy and wetting properties.

The material that fulfilled all requirements was the NILcure® JR5 material decribed above. Figure 6 shows SEM images of the full sequence from master over stamp to imprint using the NILcure® JR5 nanoimprint material. The left images show the master structure, the center images the stamp, which was replicated from the master, and to the very right the imprint is shown, which was made using the stamp. It can be seen that the undercut features are nicely replicated. Looking at the master structure in more detail, a slight tapering of the T crossbar can be noticed, which is indicated by the dashed line in the second row of images in Figure 6. The same tapering can be observed in the SEM cross-section of the stamp (top-down) and the imprint as well. This shows that in the whole process chain the replication of the undercut structures works very well.

We successfully used the same stamp for several imprints. Figure 7 shows SEM cross-sectional images of the first imprint (left) and of the fifth imprint (right), all made with the same stamp. This shows that the stamp remains intact during the imprinting process, which is of course a necessary prerequisite for a nanoimprint process. The difference in the shape of the edges of the T-structures compared to the ones shown in Figure 6 results from the fact that a different master was used, which exhibited a slightly different geometry.

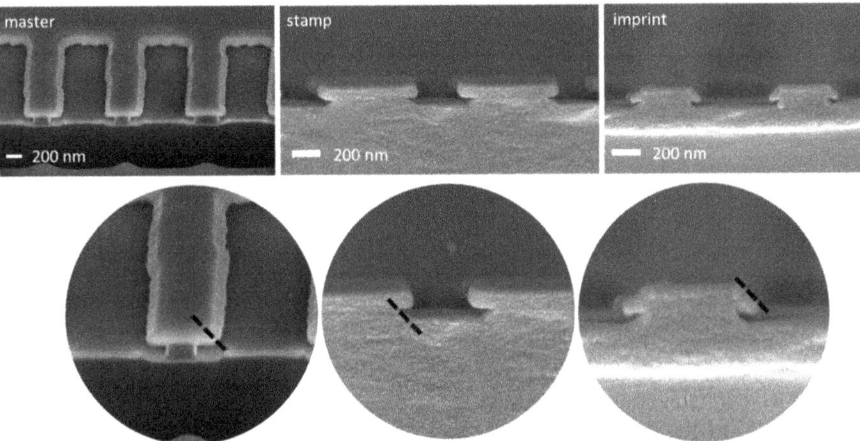

Figure 6. SEM cross-sectional images of the sample sequence from the master structure [35] (left) over the stamp in NILCure® JR5 (center) and the imprinted structures in NILcure® JR5 (right). The second row shows close-ups from the images in the upper row.

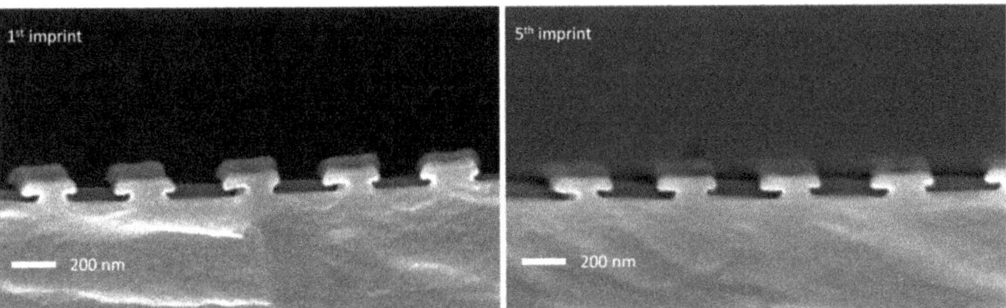

Figure 7. SEM cross-sectional images of the first imprint from (left) the master structure and the fifth imprint (right) using the same stamp.

3.2. Mulitlayer Undercut Structues

In addition, the multilayer structures could also be replicated very well with the described procedure. Figure 8 shows the sequence from the master over stamp to the imprint for the multilayer structures. Especially in the images of the stamp and the imprint, it can be seen that the tree branches partially stick together, and this can be attributed to the flexibility of the imprint in connection with the small size of the structures. It cannot be excluded that this is also an artifact of SEM imaging or sample preparation. Nevertheless, it can be seen that the nanoimprint replication of such complex features is possible.

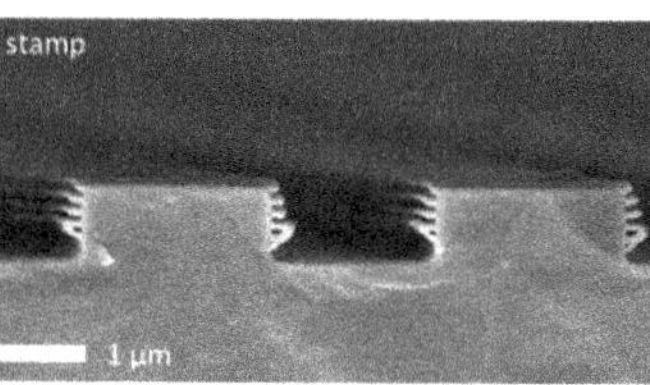
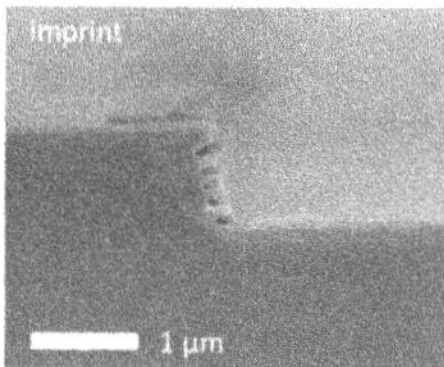

Figure 8. SEM cross-sectional images of the multilayer structure sequence from the master structure in Si (left) over the stamp in NILCure® JR5 (center) and the imprinted structures in NILcure® JR5 (right).

4. Discussion

Taking the Morpho butterfly structure as an example, we could show that nanoimprinting is capable of replicating complex undercut nanostructures. For such a process to work, it is crucial to find a material which exhibits the right wetting properties as well as suitable mechanical properties. This broadens the potential of nanoimprinting significantly and opens up the possibility for a cost effective single-step replication of, e.g., bioinspired Morpho-blue nanostructures. We tested several material combinations to arrive at a process–material combination that reliably allowed for the replication of such structures. Although all the results shown in this paper were obtained using manual nanoimprinting, we designed our processes in such a way that the transfer to roller-based processes [25,26] is possible. The roller-based nanoimprint processes will also mimic the peeling-like separation of the stamp from the imprint, which was used successfully in our experiments.

Further research will be necessary to see how the material formulation and material properties and the dimensions of the undercut nanostructures are related, i.e., if the same material can be used for significantly larger or smaller structures. The research presented here is, in our opinion, a first important step since it shows for the first time that the replication of undercut nanostructures is feasible using UV-based nanoimprint lithography.

Regarding the Morpho butterfly structure, it will be interesting to introduce the necessary randomness in the master design to faithfully reproduce the optical effects. Using nanoimprinting, it then should be possible to replicate these structures in a single imprinting step in a cost efficient way. This will be facilitated in more automated processes which will also reduce defectivity.

Depending on the application, it might be interesting or necessary to protect the imprinted nanostructures, which would require a material with a proper refractive index contrast to the imprinting resin—a topic which we did not address so far.

Furthermore, we hope to come across many more interesting nanostructures, for all types of applications, which could be enabled using this or similar processes.

Author Contributions: Conceptualization, M.M.; investigation, S.R., D.N., A.M., M.R.B., S.K., P.T., H.D.W., A.P., M.M.; writing—original draft preparation, M.M.; writing—review and editing, S.R., D.N., S.K., A.M.; project administration, M.M.; funding acquisition, M.M., D.F. All authors have read and agreed to the published version of the manuscript.

Funding: This research was funded by the rollerNIL project (FFG, grant 843639).

Data Availability Statement: The data presented in this study are available on request from the corresponding author.

Acknowledgments: The authors acknowledge administrative support by Hannes Fachberger and Eva Mühlberghuber (PROFACTOR), as well as highly interesting and very helpful discussions with Jan Stensborg and Leif Yde (Stensborg A/S), Christof Neuhauser (D. Swarovski KG), Stephan Trassl (Hueck Folien GmbH), Jan Klein (micro resist technology GmbH), Uwe Richter (SENTECH Instruments GmbH) as well as Martin Panholzer, Jean-Pierre Perin and Kurt Hingerl (Johannes Kepler University Linz).

Conflicts of Interest: The authors declare no conflict of interest.

References

1. Mühlberger, M.; Rohn, M.; Danzberger, J.; Sonntag, E.; Rank, A.; Schumm, L.; Kirchner, R.; Forsich, C.; Gorb, S.; Einwögerer, B.; et al. UV-NIL fabricated bio-inspired inlays for injection molding to influence the friction behavior of ceramic surfaces. *Microelectron. Eng.* **2015**, *141*, 140–144. [CrossRef]
2. Baum, M.J.; Heepe, L.; Fadeeva, E.; Gorb, S.N. Dry friction of microstructured polymer surfaces inspired by snake skin. *Beilstein J. Nanotechnol.* **2014**, *5*, 1091–1103. [CrossRef] [PubMed]
3. Filippov, A.E.; Gorb, S.N. Modelling of the frictional behaviour of the snake skin covered by anisotropic surface nanostructures. *Sci. Rep.* **2016**, *6*, 23539. [CrossRef] [PubMed]
4. Comanns, P.; Effertz, C.; Hischen, F.; Staudt, K.; Böhme, W.; Baumgartner, W. Moisture harvesting and water transport through specialized micro-structures on the integument of lizards. *Beilstein J. Nanotechnol.* **2011**, *2*, 204–214. [CrossRef]
5. Buchberger, G.; Kogler, A.; Weth, A.; Baumgartner, R.; Comanns, P.; Bauer, S.; Baumgartner, W. "Fluidic diode" for passive unidirectional liquid transport bioinspired by the spermathecae of fleas. *J. Bionic Eng.* **2018**, *15*, 42–56. [CrossRef]
6. Ivanova, E.P.; Hasan, J.; Webb, H.K.; Truong, V.K.; Watson, G.S.; Watson, J.A.; Baulin, V.A.; Pogodin, S.; Wang, J.Y.; Tobin, M.J.; et al. Natural Bactericidal Surfaces: Mechanical Rupture of Pseudomonas aeruginosa Cells by Cicada Wings. *Small* **2012**, *8*, 2489–2494. [CrossRef]
7. Elbourne, A.; Crawford, R.J.; Ivanova, E.P. Nano-structured antimicrobial surfaces: From nature to synthetic analogues. *J. Colloid Interface Sci.* **2017**, *508*, 603–616. [CrossRef]
8. Wilson, S.; Hutley, M. The Optical Properties of 'Moth Eye' Antireflection Surfaces. *Opt. Acta Int. J. Opt.* **1982**, *29*, 993–1009. [CrossRef]
9. Sun, J.; Wang, X.; Wu, J.; Jiang, C.; Shen, J.; Cooper, M.A.; Zheng, X.; Liu, Y.; Yang, Z.; Wu, D. Biomimetic Moth-eye Nanofabrication: Enhanced Antireflection with Superior Self-cleaning Characteristic. *Sci. Rep.* **2018**, *8*, 1–10. [CrossRef]
10. Jacobo-Martín, A.; Rueda, M.; Hernández, J.J.; Navarro-Baena, I.; Monclús, M.A.; Molina-Aldareguia, J.M.; Rodríguez, I. Bioinspired antireflective flexible films with optimized mechanical resistance fabricated by roll to roll thermal nanoimprint. *Sci. Rep.* **2021**, *11*, 1–15. [CrossRef]
11. McCoy, D.E.; McCoy, V.E.; Mandsberg, N.K.; Shneidman, A.V.; Aizenberg, J.; Prum, R.O.; Haig, D. Structurally assisted super black in colourful peacock spiders. *Proc. R. Soc. B Boil. Sci.* **2019**, *286*, 20190589. [CrossRef]
12. Butt, H.; Yetisen, A.K.; Mistry, D.; Khan, S.A.; Hassan, M.U.; Yun, S.H. MorphoButterfly-Inspired Nanostructures. *Adv. Opt. Mater.* **2016**, *4*, 497–504. [CrossRef]
13. Mason, C.W. Structural Colors in Feathers. I. *J. Phys. Chem.* **1923**, *27*, 201–251. [CrossRef]
14. Mason, C.W. Structural Colors in Insects. I. *J. Phys. Chem.* **1926**, *30*, 383–395. [CrossRef]
15. Forberich, K.; Dennler, G.; Scharber, M.C.; Hingerl, K.; Fromherz, T.; Brabec, C.J. Performance improvement of organic solar cells with moth eye anti-reflection coating. *Thin Solid Films* **2008**, *516*, 7167–7170. [CrossRef]
16. Han, K.-S.; Shin, J.-H.; Yoon, W.-Y.; Lee, H. Enhanced performance of solar cells with anti-reflection layer fabricated by nano-imprint lithography. *Sol. Energy Mater. Sol. Cells* **2011**, *95*, 288–291. [CrossRef]
17. Li, Q.; Zeng, Q.; Shi, L.; Zhang, X.; Zhang, K.-Q. Bio-inspired sensors based on photonic structures of Morpho butterfly wings: A review. *J. Mater. Chem. C* **2016**, *4*, 1752–1763. [CrossRef]
18. Chou, S.Y.; Krauss, P.R.; Renstrom, P.J. Nanoimprint lithography. *J. Vac. Sci. Technol. B Microelectron. Nanometer Struct.* **1996**, *14*, 4129. [CrossRef]
19. Haisma, J. Mold-assisted nanolithography: A process for reliable pattern replication. *J. Vac. Sci. Technol. B Microelectron. Nanometer Struct.* **1996**, *14*, 4124. [CrossRef]
20. Schift, H. Nanoimprint lithography: An old story in modern times? A review. *J. Vac. Sci. Technol. B Microelectron. Nanometer Struct.* **2008**, *26*, 458. [CrossRef]
21. Hua, F.; Sun, Y.; Gaur, A.; Meitl, M.A.; Bilhaut, L.; Rotkina, L.; Wang, J.; Geil, P.; Shim, M.; Rogers, J.A.; et al. Polymer Imprint Lithography with Molecular-Scale Resolution. *Nano Lett.* **2004**, *4*, 2467–2471. [CrossRef]
22. Müehlberger, M.; Boehm, M.; Bergmair, I.; Chouiki, M.; Schoeftner, R.; Kreindl, G.; Kast, M.; Treiblmayr, D.; Glinsner, T.; Miller, R.; et al. Nanoimprint lithography from CHARPAN Tool exposed master stamps with 12.5 nm hp. *Microelectron. Eng.* **2011**, *88*, 2070–2073. [CrossRef]
23. Reboud, V.; Obieta, I.; Bilbao, L.; Sáez-Martínez, V.; Brun, M.; Laulagnet, F.; Landis, S. Imprinted hydrogels for tunable hemispherical microlenses. *Microelectron. Eng.* **2013**, *111*, 189–192. [CrossRef]

24. Landis, S.; Reboud, V.; Enot, T.; Vizioz, C. Three dimensional on 300mm wafer scale nano imprint lithography processes. *Microelectron. Eng.* **2013**, *110*, 198–203. [CrossRef]

25. Leitgeb, M.; Nees, D.; Ruttloff, S.; Palfinger, U.; Götz, J.; Liska, R.; Belegratis, M.R.; Stadlober, B. Multilength Scale Patterning of Functional Layers by Roll-to-Roll Ultraviolet-Light-Assisted Nanoimprint Lithography. *ACS Nano* **2016**, *10*, 4926–4941. [CrossRef]

26. Yde, L.; Lindvold, L.; Stensborg, J.; Voglhuber, T.; Außerhuber, H.; Wögerer, S.; Fischinger, T.; Mühlberger, M.; Hackl, W. Roll-to-Plate UV-Nanoimprinting for Micro and Nano-Optics. Available online: http://www.nntconf.org/submissions/rollerNIL%20NNT%20R2P%20v1.pdf (accessed on 28 August 2020).

27. Moharana, A.R.; Außerhuber, H.M.; Mitteramskogler, T.; Haslinger, M.J.; Mühlberger, M.M. Multilayer Nanoimprinting to Create Hierarchical Stamp Masters for Nanoimprinting of Optical Micro- and Nanostructures. *Coatings* **2020**, *10*, 301. [CrossRef]

28. Wanzenboeck, H.D.; Waid, S.; Bertagnolli, E.; Muehlberger, M.; Bergmair, I.; Schoeftner, R. Nanoimprint lithography stamp modification utilizing focused ion beams. *J. Vac. Sci. Technol. B Microelectron. Nanometer Struct.* **2009**, *27*, 2679. [CrossRef]

29. Möllenbeck, S.; Bogdanski, N.; Wissen, M.; Scheer, H.-C.; Zajadacz, J.; Zimmer, K. Investigation of the separation of 3D-structures with undercuts. *Microelectron. Eng.* **2007**, *84*, 1007–1010. [CrossRef]

30. Möllenbeck, S.; Bogdanski, N.; Scheer, H.-C.; Zajadacz, J.; Zimmer, K. Moulding of arrowhead structures. *Microelectron. Eng.* **2009**, *86*, 608–610. [CrossRef]

31. Saito, A.; Miyamura, Y.; Ishikawa, Y.; Murase, J.; Akai-Kasaya, M.; Kuwahara, Y. Reproduction, mass production, and control of the Morpho butterfly's blue. *SPIE MOEMS-MEMS Micro Nanofabrication* **2009**, *7205*, 720506. [CrossRef]

32. Tabata, H.; Kumazawa, K.; Funakawa, M.; Takimoto, J.-I.; Akimoto, M. Microstructures and Optical Properties of Scales of Butterfly Wings. *Opt. Rev.* **1996**, *3*, 139–145. [CrossRef]

33. Saito, A.; Ishibashi, K.; Ohga, J.; Kuwahara, Y. *Morpho-Colored Flexible Film Fabricated by Simple Mass-Production Method*; Knez, M., Lakhtakia, A., Martín-Palma, R.J., Eds.; International Society for Optics and Photonics: Portland, OR, USA, 2017.

34. Saito, A.; Ishikawa, Y.; Miyamura, Y.; Akai-Kasaya, M.; Kuwahara, Y. Optimization of reproduced Morpho -blue coloration. *Opt. East* **2007**, *2007*, 676706.

35. Taus, P.; Prinz, A.; Wanzenboeck, H.D.; Schuller, P.; Tsenov, A.; Schinnerl, M.; Shawrav, M.M.; Haslinger, M.; Muehlberger, M. Mastering of NIL Stamps with Undercut T-Shaped Features from Single Layer to Multilayer Stamps. *Nanomaterials* **2021**, submitted.

36. Choi, S.-J.; Kim, H.N.; Bae, W.G.; Suh, K.-Y. Modulus- and surface energy-tunable ultraviolet-curable polyurethane acrylate: Properties and applications. *J. Mater. Chem.* **2011**, *21*, 14325–14335. [CrossRef]

37. Yoo, P.J.; Choi, S.-J.; Kim, J.H.; Suh, D.; Baek, S.J.; Kim, A.T.W.; Lee, H.H. Unconventional Patterning with A Modulus-Tunable Mold: From Imprinting to Microcontact Printing. *Chem. Mater.* **2004**, *16*, 5000–5005. [CrossRef]

38. 3300 Series Universal Testing Systems up to 50, KN. Available online: http://www.instron.us/en-us/products/testing-systems/universal-testing-systems/low-force-universal-testing-systems/3300-series-universal-testing-systems-up-to-50-kn (accessed on 23 March 2021).

39. Jastrzebski, Z.D. *Nature and Properties of Engineering Materials*; John Wiley & Sons: Hoboken, NJ, USA, 1977; ISBN 0-471-02859-2.

40. Stauffer, C.E. The Measurement of Surface Tension by the Pendant Drop Technique. *J. Phys. Chem.* **1965**, *69*, 1933–1938. [CrossRef]

41. Pendant Drop. Available online: http://www.kruss-scientific.com/en/know-how/glossary/pendant-drop (accessed on 23 March 2021).

42. Krüss DROP SHAPE ANALYZER—DSA100. Available online: https://www.kruss-scientific.com/files/kruss-techdata-dsa100-en.pdf (accessed on 23 March 2021).

43. Schmid, H.; Michel, B. Siloxane Polymers for High-Resolution, High-Accuracy Soft Lithography. *Macromolecules* **2000**, *33*, 3042–3049. [CrossRef]

44. Fluorolink®MD700. Available online: https://www.solvay.com/en/product/fluorolink-md700 (accessed on 23 March 2021).

45. OrmoComp®—Microresist. Available online: https://www.microresist.de/en/produkt/ormocomp/ (accessed on 23 March 2021).

46. Profactor. ANTI STICKING LAYER BGL-GZ-83. Available online: https://www.profactor.at/en/solutions/coatings/ (accessed on 19 February 2021).

47. Mühlberger, M.; Bergmair, I.; Klukowska, A.; Kolander, A.; Leichtfried, H.; Platzgummer, E.; Loeschner, H.; Ebm, C.; Grützner, G.; Schöftner, R. UV-NIL with working stamps made from Ormostamp. *Microelectron. Eng.* **2009**, *86*, 691–693. [CrossRef]

48. Biostep. Available online: https://biostep.de/ (accessed on 24 February 2021).

49. Profactor. HMNP-12—Adhesion Layer for Nanoimprint Materials. Available online: https://www.profactor.at/en/solutions/coatings/ (accessed on 19 February 2021).

50. Coveme Treated and Stabilized Films—Kemafoil HSPL/HSPL W. Available online: https://www.coveme.com/hspl-w-high-surface-tension-film/ (accessed on 19 February 2021).

51. COVEME BIOMEDICAL—Films for Biosensors in Near Patient Diagnostic. Available online: https://www.coveme.com/files/documenti/divisioni-brochure/brochure_biomedical_web.pdf (accessed on 19 February 2021).

52. MELINEX®ST506. Available online: http://52.167.112.238/FilmEnterprise/Datasheet.asp?ID=271&Version=US (accessed on 19 February 2021).

53. Fisher Scientific. Thermo Scientific™ SuperFrost Ultra Plus™ Adhäsionsobjektträger. Available online: https://www.fishersci.de/shop/products/superfrost-ultra-plus-adhesion-slides-3/10417002 (accessed on 19 February 2021).

54. Boeckeler Instruments. Available online: https://www.boeckeler.com/company (accessed on 19 February 2021).

 *nanomaterials*

Article

Optical Polymer Waveguides Fabricated by Roll-to-Plate Nanoimprinting Technique

Vaclav Prajzler [1,*], Vaclav Chlupaty [1], Pavel Kulha [1,2,*], Milos Neruda [1], Sonja Kopp [2] and Michael Mühlberger [2]

[1] Department of Microelectronics, Faculty of Electrical Engineering, Czech Technical University in Prague, Technicka 2, 168 27 Prague, Czech Republic; chlupvac@fel.cvut.cz (V.C.); nerudmil@fel.cvut.cz (M.N.)

[2] PROFACTOR GmbH, Im Stadtgut D1, A-4407 Steyr-Gleink, Austria; Sonja.Kopp@profactor.at (S.K.); michael.muehlberger@profactor.at (M.M.)

* Correspondence: vaclav.prajzler@fel.cvut.cz (V.P.); pavel.kulha@profactor.at (P.K.); Tel.: +420-22435-2338 (V.P.); +43-7252885-431 (P.K.)

Abstract: The paper reports on the properties of UV-curable inorganic-organic hybrid polymer multi-mode optical channel waveguides fabricated by roll-to-plate (R2P) nanoimprinting. We measured transmission spectra, refractive indices of the applied polymer materials, and optimized the R2P fabrication process. Optical losses of the waveguides were measured by the cut-back method at wavelengths of 532, 650, 850, 1310, and 1550 nm. The lowest optical losses were measured at 850 nm and the lowest average value was 0.19 dB/cm, and optical losses at 1310 nm were 0.42 dB/cm and 0.25 dB/cm at 650 nm respectively. The study has demonstrated that nanoimprinting has great potential for the implementation of optical polymer waveguides not only for optical interconnection applications.

Keywords: optical planar waveguides; roll-to-plate R2P nanoimprinting; UV-curable polymers; inorganic-organic hybrid polymer; optical losses

Citation: Prajzler, V.; Chlupaty, V.; Kulha, P.; Neruda, M.; Kopp, S.; Mühlberger, M. Optical Polymer Waveguides Fabricated by Roll-to-Plate Nanoimprinting Technique. *Nanomaterials* **2021**, *11*, 724. https://doi.org/10.3390/nano11030724

Academic Editor: Andrea Chiappini

Received: 24 February 2021
Accepted: 8 March 2021
Published: 13 March 2021

Publisher's Note: MDPI stays neutral with regard to jurisdictional claims in published maps and institutional affiliations.

1. Introduction

Integrated optical and photonic devices are playing an increasingly important role in optical communication networks, optical interconnections, optical data centers and the application of optical sensors [1]. The importance of these integrated optical devices grows even more due to the rapid widespread communication devices for smart cities data communications using internet applications in the Fiber-to-the-Home (FTTH) and Internet of Things (IoT) systems. Optical planar waveguides are the basic building blocks for the implementation of these optics and photonic structures and devices.

Up to now, various materials, including semiconductors such as silicon, silicon nitride, indium phosphide, III–V compound, silica, or optical crystals, e.g., lithium niobate, lithium tantalate, and rubidium titanyl phosphate, have been used for the fabrication of the optical waveguides [2]. Highly integrated optics and photonic devices fabricated from polymers have been subject to intense research in recent years [3]. The advantages of polymer materials compared above mention materials are easier fabrication processes and that lead to developed optics devices with significantly lower material and production costs.

Therefore, in the last decades, new types of polymers for optics and photonics applications were developed in many laboratories worldwide and some of them are commercially available [4,5]. These polymers include siloxane LIGHTLINK™ XP-6701A core, LIGHTLINK™ XH-100145 clad [6], UV-curable epoxy polymers Su-8, EpoCore/EpoClad [7–9], benzocyclobutene (Dow Chemical, Midland, MI, USA) [10], ZPU resin and polymers (ChemOptics Inc., Daejeon, South Korea) [11], inorganic–organic hybrid polymers OrmoClear®FX (micro resist technology GmbH), SUNCONNECT (Nissan Chemical Ltd., Tokyo, Japan) [12,13], UV exposure optical elastomer OE-4140 core, OE-4141 cladding (Dow Corning, Midland,

MI, USA) [14], Truemode Backplane Polymer (Exxelis, Ltd., Washington, DC, USA) [15], polydimethylsiloxane Sylgard 184, LS-6943 [16,17] and etc. [4,5].

These new polymers have unique and excellent optical properties such as low optical losses at operating wavelengths (including infra-red spectrum), well-controlled and tuneable refractive indices, thermal and chemical resistance, mechanical, environmental stability and environmental-friendly fabrication processes etc. Optical planar waveguides are basic building blocks for the realization of optics and photonic devices and several different techniques for the fabrication of polymer optics waveguides devices have been reported. These fabrication methods include mask photolithographic technology and following wet etching process [9], photo-resist patterning combined with reactive ion etching [18], two-photon-polymerization [19], laser direct writing [8], electron beam writing [20], flexographic and inkjet printing [21], hot embossing process [22], photo-bleaching [23], etc. These methods involve many processing steps and can lead to long fabrication times and low yield. Therefore, technologies such as the stamping methods [24] are studied for mass production. These methods include roll-to-roll (R2R) nanoimprint lithography processes [25,26] and also roll-to-plate nanoimprinting. Theses roller-based technologies provides an opportunity to use polymers, which are leading material candidates for applications requiring inexpensive and mass productions [26]. These technologies are used for the fabrication of flexible electronics [27]. Further, possibilities for use in optics and photonics applications are now being studied.

In this work, we demonstrate the fabrication of the optical waveguides using a Roll-to-Plate (R2P) imprinting process. R2P imprint lithography is an imprinting process that employs a roller-mounded stamp (imprinting plate) and a rigid surface plate, where the substrate is mounted. The core of our R2P fabrication unit is a transparent cylinder that houses a UV-source in its center [28]. A distinct advantage of roller-based imprinting is the peeling-like separation process of imprinting plate and substrate. This facilitates the replication of complex structures as well as the imprinting on large areas. Roller-based imprinting has been proposed soon after the invention of imprinting and has been developed further in various variants like roll-to-roll or roll-to-plate [25,26]. In this paper, we focused on the R2P imprint replication of microscale structures. We optimized the fabrication procedure to the dimensions of the core waveguides 50×50 μm (width $\times$ height). These dimensions were used due to the standard dimensions of multi-mode optical waveguides used for optics communications. In many cases, it is more challenging to nanoimprint micro-sized features as compared to nano-sized features since the material, that has to be displaced during the imprinting process, has to be moved over larger distances (several tens of μm) as compared to nano-sized features, where the displacement takes place only over several 100 nm typically. Nevertheless, imprinting has distinct advantages also for those types of applications and feature sizes, like the direct patterning of functional materials and large area patterning as can be seen below.

2. Materials and Methods

Optical waveguides were fabricated using UV photopolymer Lumogen OVD Varnish 311 (BASF) for the cladding layer and UV-curable inorganic-organic hybrid polymer OrmoClear®FX (micro resist technology GmbH, Berlin, Germany) were used for the core layer. Properties of the UV photopolymer Lumogen OVD Varnish 311 (BASF) provided by the supplier are as follows: density 1.1 g/mL (22 °C), viscosity Brookfield (12 rpm) 75–120 mPa·s, the refractive index measured by ellipsometry 623.8 nm (He-Ne) $n = 1.507$. Properties of the OrmoClear®FX polymer are as follows: viscosity 1.5 ±0.3 Pa·s, the refractive index 1.555 (589 nm, exposure).

A nickel master mold was used for the fabrication of the polydimethylsiloxane (PDMS) master stamp. The PDMS master was made from Sylgard 184 (The Dow Chemical Company, purchased at ELCHEMCo Ltd., Zruč nad Sázavou, Czechia). A HoloPrint® uniA6 DT nano imprinter (Stensborg A/S) was used for roll-to-plate (R2P) nanoimprinting (NIL) process.

The transmission spectra were measured with a UV-VIS-NIR spectrometer (UV-3600 Shimadzu, Shimadzu Deutschland GmbH, Duisburg, Germany) in the spectral range of 250–1750 nm. The refractive indices of the samples were measured by dark mode spectroscopy using the Metricon 2010/M prism-coupler system and the measurement was done at six wavelengths 532.0, 654.2, 846.4, 1308.2, 1549.1 and 1652.1 nm and set for transverse-electric (TE) polarisation. We used prism #200-P-4a with refractive index $n = 2.1558$ ($\lambda = 632.8$ nm) and the applied prism enabled the measurement of the range of effective refractive indices from 1.2 to 2.02 at wavelength $\lambda = 632.8$ nm.

The optical/geometrical inspections of the fabricated optical channel waveguides were carried out by an optical digital camera ARTCAMI equipped with optical head ZOOM Optics (Olympus Czech Group Ltd., Prague, Czech Republic) and the software QUICK-FOTO (the Version 3.0, PROMICRA, Prague, Czech Republic) for the control of waveguides dimensions. The dimension of the fabricated stamps and optical waveguide channels were also inspected by the KEYENCE VHX-5000 microscope (KEYENCE INTERNATIONAL, Mechelen, Belgium).

The optical properties of the optical channel waveguides were determined using the cut-back method at wavelengths: 532 nm (laser Nd:YVO$_4$), 650 nm (laser Safibra OFLS-5-FP-650), 850 nm (laser Safibra OFLS-6-LD-850), 1310 nm (laser Safibra OFLS-6CH, SLED-1310) and 1550 nm (laser Safibra OFLS-5-DFB-1550). The method is described in more detail below. The input light was coupled into the channel waveguide using a 50 μm core multi-mode fiber; a larger-core (62.5 μm) fibre was utilized as the output. The input/output fibres were precisely aligned to the channel waveguides by using high-precision 3-axis stages on the optical bench; the output light power intensity was detected by a Thorlabs PM200 optical power meter. A silicon detector (Thorlabs S151C) was used for the measurement at wavelengths of 532, 650, and 850 nm and an Indium gallium arsenide detector (Thorlabs S155C) was utilized for the measurement at 1310 and 1550 nm. Optical losses α were calculated using the following equation:

$$\alpha = \frac{10 \cdot \log\frac{P_1(W)}{P_2(W)}}{l_1 - l_2(cm)},$$

(1)

where l_1, l_2 are the lengths of the channel waveguides and P_1, P_2 are output optical powers before and after cutting the waveguide, respectively. The accuracy of the optical measurement set-up is estimated to $\pm 5\%$.

The fabrication process was as follows: The nickel master mold was used for the fabrication of the PDMS elastomer stamp. This nickel negative mold was made by the galvanoplastic process of a photoresist master produced by the photolithographic method. The mold was 8 cm long and had 12 channels with dimensions of 50×50 μm and 250 μm pitch between channels. PDMS stamps were fabricated from Sylgard 184 elastomer and elastomer was prepared by mixing the A and B agents in the ratio 10:1 and the mixture was stirred and then evacuated in a desiccator for 60 min. Then, the elastomer was poured onto the nickel mold and then the hardening process was done in the oven at 125 °C for 20 min (Figure 1a).

After cooling, the PDMS stamp was carefully torn off from the nickel mold (Figure 1b) and it was treated by separator SP-3 (ELCHEMCo Ltd., Zruč nad Sázavou, Czechia). Then the PDMS stamp was fixed on the cylinder of the R2P machine. After that, the polymer Lumogen OVD Varnish 311 cladding layer with a thickness of 500 μm was deposited onto the glass substrate by using the doctor blade technique (Figure 1c). The R2P machine was set properly before the start of the imprinting process. The most important parameters are the UV light intensity and the position of the cylinder height, and the imprinting speed. Because the cylinder height settings depend on the thickness of stamp and substrate, they have to be determined individually for each stamp/substrate thickness combination. The UV-light source in the R2P NIL tool uses 395 nm LEDs. After setting all parameters of the R2P machine, the PDMS stamp was imprinted into Varnish 311 UV photopolymer

(Figure 1d,e). After that UV-curable inorganic-organic hybrid polymer OrmoClear®FX was deposited by doctor blading into the U-grooves from the Varnish substrate (Figure 1f). The OrmoClear®FX core layer was hardened by UV light @ 365 nm for 60 s (dose 100 mW/cm^2). Next, a Varnish 311 UV photopolymer cover cladding layer was also fabricated by using the doctor blading technique (Figure 1g) and finally, the waveguide structure was torn off from the glass substrate (Figure 1h). Before separating the waveguides structure from the glass substrate, the Varnish 311 cover layer was hardened by a Mercury lamp (dose 1500 mJ/cm^2). This process of the post UV curing finished hardening all other polymer layers (Lumogen OVD Varnish 311 substrate and an upper cladding layer, OrmoClear®FX core) which did not cure completely by the short bandwidth UV-light. This allowed the layers to bond together to create one integrated functional structure before the final UV curing step.

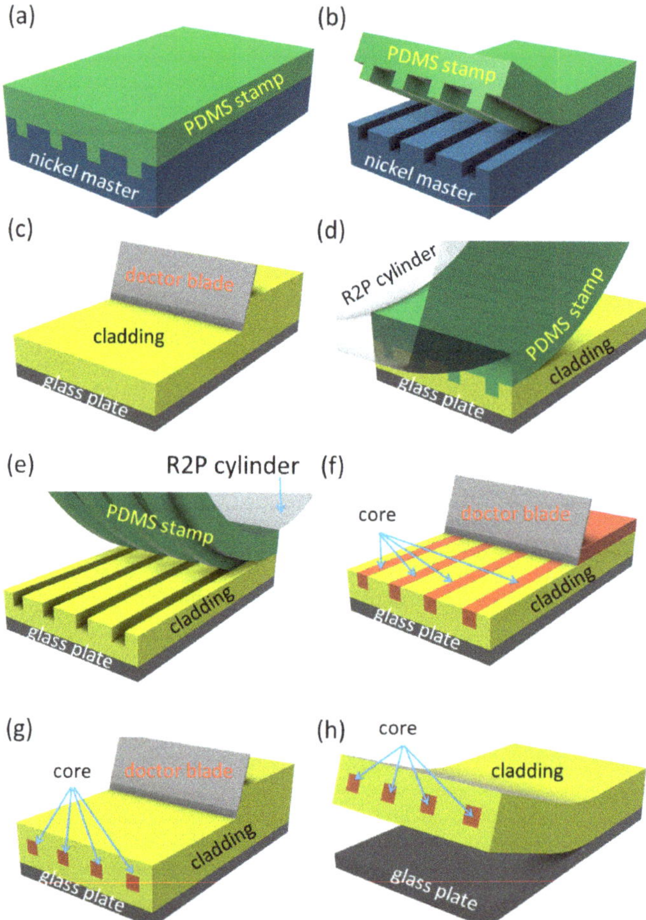

Figure 1. Fabrication process of the multimode optical waveguides using Roll-to-Plate imprinting, (**a**) fabrication PDMS (Sylgard 184) stamp layer, (**b**) separating stamp and Ni-mold, (**c**) fabrication of Varnish 311 UV layer, (**d,e**) fabrication of the U-groove into Varnish 311 UV substrate layer by R2P process, (**f**) fabrication of the OrmoClear®FX core layer into the U-groove Varnish 311 substrate, (**g**) fabrication of Varnish 311 UV cover cladding layer, (**h**) separating of waveguide structure from the glass substrate.

Finally, the optical waveguide end-facets for input/output optical fiber coupling were formed with a scalpel and polished. We used a three-step polishing process using polished (lapping) sheets with grit sizes of 3, 1, and 0.2 µm. This procedure reduced the length of the waveguides from 8 cm to approximately $5 \div 7$ cm.

2.1. R2P Technology—Optimization Process

One part of our research was also to optimize the fabrication procedure to achieve optical waveguides with dimensions 50×50 µm with pitch 250 µm and its homogeneity in the whole length of the waveguides. For the realization of optical channel waveguides with precision dimensions, it is important to fabricate a high-quality PDMS stamp, which is then copied into the UV polymer substrate layer using R2P imprinting. We fabricated this stamp by using a nickel master mold. An example of the fabricated stamp is shown in Figure 2. Figure 2a shows a picture whole PDMS stamp (length 8 cm) and Figure 2b shows a detailed KEYENCE VHX-5000 microscope picture of the one ridge channel. We expected to create channels with dimensions 50×50 µm with accuracy ± 5 µm. This picture proved the good quality of the stamp with sharp edges with required dimensions.

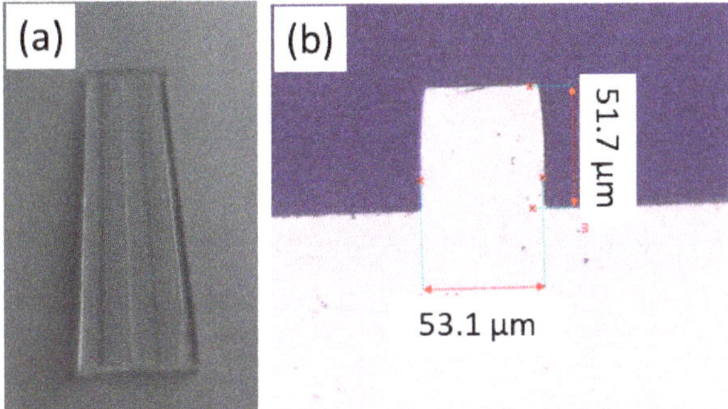

Figure 2. Picture of the Sylgard 184 stamp (**a**) camera picture whole stamp sample (total length of the PDMS stamp 8 cm), (**b**) microscope detail picture of the 50×50 µm channel.

The dimensional precision of the fabricated optical channels waveguides depends on the distance between and glass plate and the R2P cylinder (see Figure 3a), which translates to an imprinting pressure, the speed of the plate moving under the R2P cylinder, and the UV curing dose applied not only during the R2P process, but also as post-processing UV curing.

In the case of the distance between the R2P cylinder and glass plate with the Lumogen OVD Varnish 311 cladding polymer was too high, the PDMS stamp was not imprinted into Varnish 311 polymer correctly and also the air bubbles were observed in the polymer layer. On the other hand, if the distance between the R2P cylinder and glass plate was too low the PDMS stamp was deformed due to the high pressure and the shape and dimensions of the channels were also not right (see Figure 3b, picture above). The channels were also not in the same right level position (see Figure 3b, picture below). In the HoloPrint® (Dubai, United Arab Emirates) uniA6 DT, setting the roller height is done mechanically and the distance between substrate and roller can be up to 8 mm.

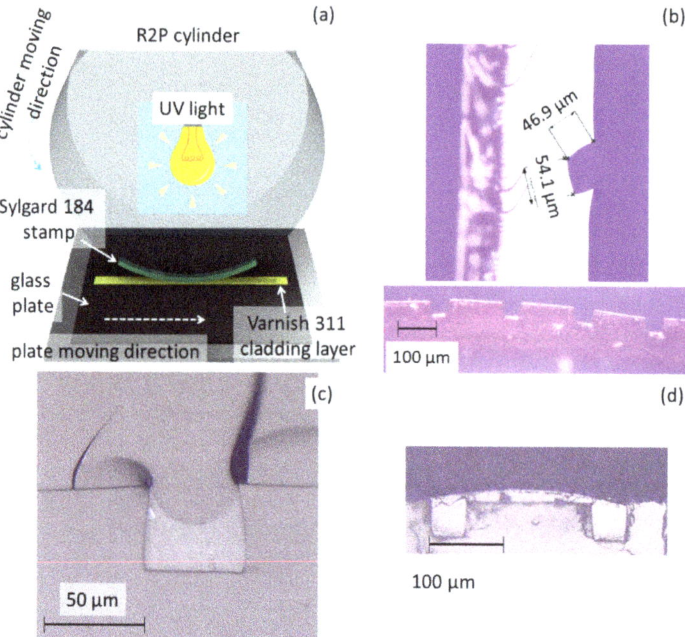

Figure 3. (**a**) Principle of the Roll-to-plate technology process. Results incorrectly set R2P process, KEYENCE microscope pictures (**b**) example of the fabricated U-groove channel in Lumogen OVD Varnish 311 by PDMS stamp, (**c**) example of the Lumogen OVD Varnish 311 U-groove fill in OrmoClear®FX core—channel is not filled correctly with the polymer core, (**d**) example of the Lumogen OVD Varnish 311 U-groove fill in OrmoClear®FX core—two channels are over-covered with the polymer core layer.

The UV curing process is the next important parameter and it depends on the exposure power of the applied UV source and the moving speed of the sample table. If the UV dose is higher than the curing process requires, then the polymer layer is hardened before the PDMS stamp was imprinted in the semi-cured material, therefore the channels have not the desired shape and dimensions. The case of lower UV dose expose caused an insufficient hardening of the material and the channels did not have again the correct shape the polymer was not solidified and could have liquid form after the process. In general, the UV intensity can be varied and a maximum dose of 100 mJ/cm^2 at 6 m/min can be set. The minimum setting corresponds to 10 mJ/cm^2 at 6 m/min or 30 mJ/cm^2 at 2 m/min. The optimization process showed that the lowest UV light dose was suitable for the Lumogen OVD Varnish 311 cladding polymer.

The next parameter which has to be optimized is the moving speed of the plate with the substrate and the polymer cladding layer. A higher moving speed results in a lower dose of UV light and thus the UV polymer is less exposed. The substrate holder moving speed also affects the quality of the original PDMS stamp shape. At high speeds, wave-like artefacts and defects in the form of bubbles can be observed. Too small UV expose doses caused an insufficient hardening of the polymer layer and the channels did not have the correct shape—the material flowed even after the curing process. The used R2P machine allowed to set the moving speed from 0–8 m/min and completed tests showed that for applied polymers a speed of 2 m/min proved to be the most suitable.

The next step was to optimize the filling process of the core polymer into the U-grooves fabricated by R2P NIL. The core layer is made of the UV-curable inorganic-organic hybrid polymer OrmoClear®FX and for filling the core polymer into the U-groove we used the doctor blade technique. For this technique is important that the stamped layer was straight flat for easy squeegeeing the core polymer layer into U-groove channels. The next important thing was to choose an appropriate blade. Therefore, various blades were tested including plastic, rubber squeegees, and iron razor blades. The blades made of plastics and rubbers were too thick and the polymer core material was partly removed from the U-groove (see Figure 3c). Harder squeegees, in turn, left excess material between the channels (see Figure 3d). The best result was obtained when wiping using a sharp razor blade, which was sufficiently flexible and adapted to the wiped surface but did not remove core polymer from the channels. Therefore, this type of blade allowed homogeneous filling-up of the U-channels.

2.2. Stamp Modification

The PDMS stamp in the function locations was made to the desired quality, but a spatial nonuniformity was formed on the left and right edges of the stamp (see Figure 4a) due to the shape of the Ni master. After the PDMS stamp was imprinted into the Lumogen OVD Varnish cladding layers, initially inequalities arose on the sides of the sample (see Figure 4b), which prevented the perfect wiping of the core layer when applying doctor blade technology and this layer remained on the cladding layer (see Figure 4c).

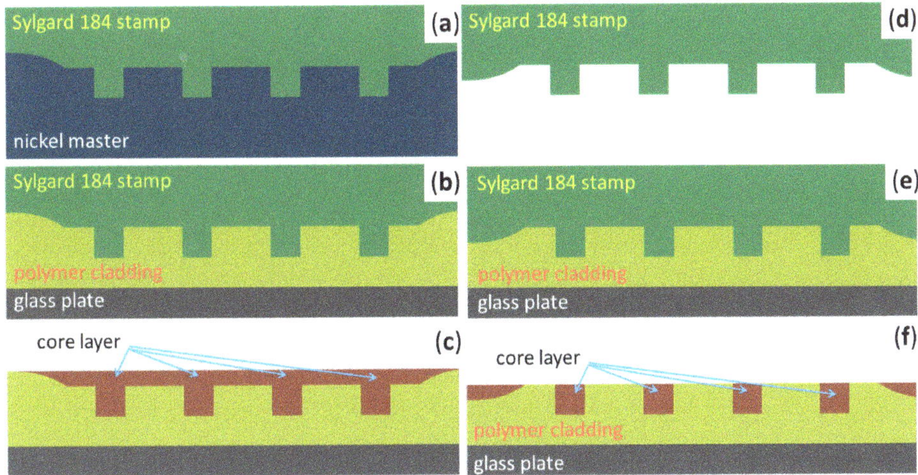

Figure 4. Illustrations of the Sylgard 184 stamp modification process. (**a**) Fabrication of the Sylgard 184 stamp using nickel master, (**b**) fabrication of the U-groove channels using Sylgard 184 stamp, (**c**) filing core polymer layer into U-groove channels, (**d**) modification of the Sylgard 184 stamp, (**e**) fabrication of the U-groove channels using modified Sylgard 184 stamp, (**f**) filing core polymer layer into U-groove channels.

Due to this unevenness (see Figure 4b), we performed a PDMS stamp correction to refine the dimension and shape of the overall stamp. The correction was made after the separation of the PDMS stamp from the master nickel mold and Sylgard 184 polymer was added by a pipette to both edges of the stamp (see Figure 4d). Then the fabrication process of the U-groove into polymer substrate was done by R2P NIL (see Figure 4e). Figure 4f schematically shows filling U-groove channels after the doctor blading process using a modified PDMS stamp.

After cooling, the PDMS stamp was carefully torn off from the nickel mold (Figure 1b), and then it was treated by separator SP-3 to improve stamp life (ELCHEMCo Ltd., Zruč nad Sázavou, Czechia). If the separator SP-3 was not applied, we observed dimensional inhomogeneity and channel wall roughness in the U-groove Lumogen OVD Varnish 311 substrate after a few fabricated samples. In the case whereby the PDMS stamp was treated with the separator, we did not observe any defect until after more than ten produced samples.

3. Results

3.1. UV-VIS-NIR Transmission Spectroscopy and Dark Mode Spectroscopy

The transmission spectra were measured with a UV-VIS-NIR spectrometer (UV-3600 Shimadzu, Shimadzu Deutschland GmbH, Duisburg, Germany) in the spectral range of 250–1750 nm. In Figure 5a, a comparison of the transmission spectra for Lumogen OVD Varnish 311 cladding layer and OrmoClear®FX core layer is given.

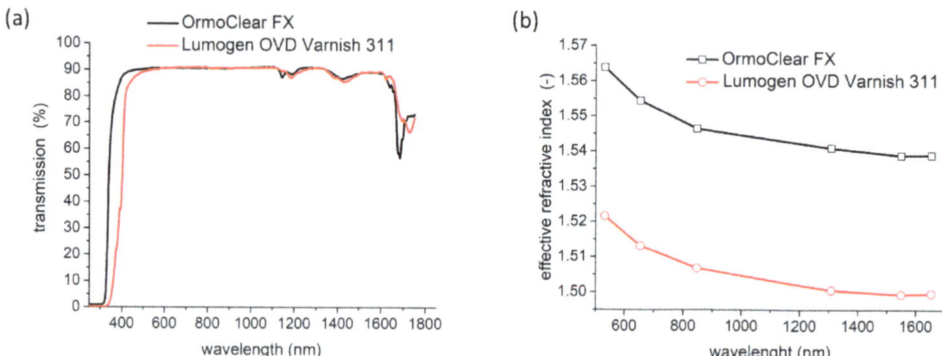

Figure 5. Properties of UV photopolymer Lumogen OVD Varnish 311 (BASF) and inorganic-organic hybrid polymer OrmoClear®FX (**a**) the transmission spectra, (**b**) the refractive indices.

The refractive indices of the used polymers were measured by dark mode spectroscopy using six wavelengths: 532.0, 654.2, 846.4, 1308.2, 1549.1 and 1652.1 nm. The principle of the method has been already described in [12] and the measured results are shown in Figure 5b.

3.2. Dimensions of the Optical Channel Waveguides

The illustration structure of the optical channel waveguide with expected dimensions is shown in Figure 6a.

The final fabricated waveguide structures have twelve waveguides with 250 μm pitch with a length of around 8 cm. The optical ARTCAMI microscope picture of the four-channel waveguides is shown in Figure 6b,c, where Figure 6b shows the cross-sectional view of the four-channel waveguides and Figure 6c shows a detail of one channel waveguide. The figures show that channels have the required dimensions of 50 × 50 μm (height × width) with the channels pitch 250 μm with required accuracy dimensions ±5 μm. The figures also proved that channels had the required quality without any observable defects. Figure 6d shows a detailed picture of the optical losses measurement set-up, where the transmission of red light (650 nm) coupled via optical fiber into the channel waveguide is clearly visible.

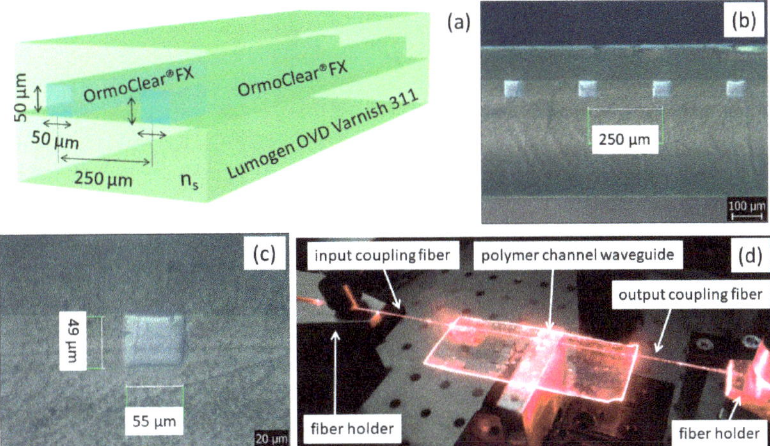

Figure 6. (**a**) Illustration of the optical channel waveguide, (**b**) the cross-sectional view of the four optical channel waveguides, (**c**) the detail view of the cross-section of a single optical channel waveguide, (**d**) photo of the measurement set-up with optical channel waveguide coupled with red light 650 nm.

3.3. Losses of the Optical Channel Waveguides

The optical losses were measured for eight samples where we measured eight optical channels in each sample (a total of 64 waveguides). The measurement started with determining the optical power P_1 coming from a laser through the input coupling fiber and passing through the whole length of the measured channel waveguide to the output fiber connected with the optical power meter. Then the sample was cut and the output power P_2 for the shortened sample was measured. The principle of the method has been previously reported in [17]. The results for the average measured values of the optical losses for eight waveguide channels are summarized in Table 1.

Table 1. Average values of the optical losses measured by the cut-back method at wavelengths 532, 650, 850, 1310 and 1550 nm.

Sample	Length (cm)	Wavelengths (nm)				
		532	650	850	1310	1550
		Optical Losses (dB/cm)				
#1	5.71	−1.21	−0.96	−0.78	−0.79	−1.52
#2	5.27	−1.01	−0.82	−0.62	−0.70	−1.33
#3	6.76	−0.50	−0.51	−0.39	−0.53	−1.17
#4	6.76	−0.40	−0.47	−0.42	−0.54	−1.23
#5	7.06	−0.55	−0.41	−0.33	−0.52	−1.50
#6	7.06	−0.31	−0.25	−0.19	−0.44	−1.37
#7	5.80	−0.42	−0.43	−0.30	−0.38	−1.13
#8	5.80	−0.64	−0.48	−0.42	−0.54	−1.49
average value	6.28	−0.63	−0.54	−0.43	−0.56	−1.34

The results show that the highest values of the optical losses were at wavelength 1550 nm and the losses were higher than 1.10 dB/cm. Three samples (#1, #2 and #5) had higher optical losses than 1.30 dB/cm, therefore OrmoClear®FX polymer is not suitable material for realization optical waveguides operating at this wavelength. The lowest values of the optical losses were obtained at wavelengths 850 nm and the lowest average value was 0.19 dB/cm (sample #6). Very good results were also found for wavelength 1310 nm where average values of the optical losses were 0.38 dB/cm (sample #7). Optical losses measurement at visible spectrum for the red light at 650 nm showed that average values of the optical losses were also low and optical losses at a green light (532 nm) were a little bit higher than on red light (650 nm). All waveguides channels have lower values of the losses less than 1 dB/cm @ 532 nm except for the first two samples (#1, #2).

The lowest values of the optical losses were measured for sample #6 and the results measurement of the optical losses for eight channels are given in Table 2.

Table 2. Optical losses for sample #6 measured by the cut-back method at wavelengths 532, 650, 850, 1310 and 1550 nm.

Wavelength (nm)	Channel No.							
	1	2	3	4	5	6	7	8
	Optical Losses (dB/cm)							
532	−0.63	−0.49	−0.49	−0.56	−0.54	−0.52	−0.52	−0.66
650	−0.48	−0.38	−0.42	−0.38	−0.34	−0.36	−0.40	−0.54
850	−0.38	−0.31	−0.31	−0.30	−0.27	−0.32	−0.30	−0.40
1310	−0.58	−0.50	−0.50	−0.50	−0.47	−0.50	−0.51	−0.61
1550	−1.54	−1.50	−1.51	−1.44	−1.40	−1.54	−1.48	−1.59

4. Discussion

Optical Properties

Transmission spectra measurement shows that UV-curable inorganic-organic hybrid polymer OrmoClear®FX core layer is transparent in the whole measured spectral range from the visible to near-infrared spectrum (250–1750 nm). The measurement showed that this polymer has suitable properties as an optical waveguide core material and this result is also in good agreement with our previously published results in [12].

The measured values of the refractive indices n presented in Figure 5b proved that inorganic-organic hybrid polymer OrmoClear®FX has higher values of the refractive indices (1.5409@ 1308.2 nm) than UV photopolymer Lumogen OVD Varnish 311 layer (1.5006@ 1308.2 nm). Therefore, OrmoClear®FX hybrid polymer is a suitable candidate for the core layer with a combination Lumogen OVD Varnish 311 cladding layer. Our measured values of the refractive indices of OrmoClear®FX were also in good agreement with our previously published data and data provided by supplier micro resist technology GmbH [12]. OrmoClear®FX hybrid polymer used for the core layer has a higher value of the refractive index (n = 1.5409@ 1308.2 nm, 1.5388@ 1549.1 nm) compared to other optical polymers for example siloxane optical LIGHTLINK™ XP-6701A polymer (n = 1.505@ 1308.2 nm) or polydimethylsiloxane elastomers Sylgard 184 (n = 1.403@ 1308.2 nm) and NuSil Technology LS-6943 (n = 1.418@ 1308.2 nm). Therefore, and because of its excellent properties such as high transparency for visible and infrared light, high thermal, mechanical and chemical stability OrmoClear®FX hybrid polymer is a suitable candidate for advanced micro and nano-optical applications. In addition, OrmoClear®FX is also from chemical composition compatible with PDMS, making this material even more interesting.

The UV photopolymer Lumogen OVD Varnish 311 is a new material and therefore published data describing properties of this material are not available at this time. The refractive index value provided by the BASF supplier measured by ellipsometry is given only for red light n = 1.507 (He-Ne, 632.8 nm). Our measurement refractive index for a red light was n = 1.5132 (654.2 nm). This value is higher by 0.006 and the difference could be caused by using different fabrication procedures with comparison to the supplier fabrication process.

The average values of the optical losses for eight samples are depicted in Table 1 and results for the best sample (#6) presented for all measured channels are presented in Table 2. Optical communication systems, which use planar optical multimode waveguides with a geometric dimension of the waveguides channels of 50 × 50 µm, are optimized for operating wavelengths of 850 or 1300 nm. These types of waveguides are used for the distribution of the data in optical interconnection, short chip-to-chip, or board-to-board communications. In the last decade, the importance of these optical waveguides is growing due to the development of the next-generation optical interconnections communications systems in data-center, high-performance computers and etc. One of the most important properties of these waveguides is low optical losses, which have to be as low as possible and must not exceed the value of 1 dB/cm at operating wavelengths. Our measured data in Table 1 proved that presented waveguides have low optical losses and can be used in an optical communication system using wavelengths 850 and 1300 nm. For special short communications systems and optical sensors applications are also used optical wavelengths in the visible spectrum and our measurement proved that our waveguides with OrmoClear®FX hybrid polymer core layer and UV photopolymer Lumogen OVD Varnish 311 fabricated by R2P technology can be used also for operation on red light 650 nm. The average values of optical losses our waveguides were −0.43 dB/cm at 850 nm, −0.56 dB/cm at 1310 nm and −0.54 dB/cm at 650 nm respectively.

We compared our results with the waveguides presented with our previous works [9,17,29] and also with other authors' [7,9,14,24,25]. We presented polymer multimode optical waveguides with the dimension of the channels 50×50 μm fabricated by the photolithography process. The waveguides were fabricated from epoxy polymer EpoCore/EpoClad and the average values of the optical losses for these waveguides were −1.85 dB/cm (650 nm), −0.60 dB/cm (850 nm) and −0.69 dB/cm (1310 nm), respectively [9].

Previously, we have already reported on flexible optical multimode elastomer polydimethyl-diphenylsiloxane channel waveguides (core layer LS-6943 NuSil, $n = 1.4184$ at 1311 nm, cladding Sylgard 184, $n = 1.4030$), where optical losses were lower than 0.45 dB/cm for four measured wavelengths 532, 650, 850 and 1310 nm) [17]. We have also presented elastomer waveguides with the same dimensions but here we used elastomer LS-6946 NuSil core layer and the average values of the optical losses were −0.76 dB/cm (650 nm), −0.59 dB/cm (850 nm), −0.56 dB/cm (1310 nm), and −1.93 dB/cm (1550 nm), respectively [29].

Bamiedakis et al. presented properties of multimode optical polymer waveguide for high-speed on-board optical interconnect fabricated from siloxane OE-4140 (core) and OE-4141 (cladding) material developed by Dow Corning and their waveguides exhibit low loss of approximately 0.04 dB/cm at 850 nm [14]. Choi et al. reported about properties of the multimode waveguide with the core size 50×50 μm and the core material of the waveguide was SU-8 and with Topas cladding. The measured propagation loss of the waveguide was 0.6 dB/cm at 850 nm [30]. Polymer-based thin-film foil waveguides fabricated by an industrial high-volume roll-to-roll embossing process were presented by Bruck et al. in [25] and they obtained propagation losses lower than 1 dB/cm for a wavelength of 633 nm. The multimode waveguides fabricated by the stamping method by Kobayashi had the propagation loss of 0.06 dB/cm at a wavelength of 850 nm [24].

There are also papers that present single mode polymer optical waveguides for example Elmogi et al., 2016 [9]. They present the properties of two types of waveguides, which were fabricated by direct-write lithography and use epoxy EpoCore_5 and siloxane XP-6701A LightLink core layers. The dimensions of the core were about 5×5 μm and for the epoxy-based system, the average propagation losses are 0.49 dB/cm and 2.23 dB/cm at 1.31 μm and 1.55 μm, respectively. For the siloxane-based waveguides, the average propagation losses were 0.34 dB/cm and 1.36 dB/cm at 1.31 μm and 1.55 μm respectively [9]. Properties of the single-mode TE_{00}-TM_{00} SU-8 rib waveguides with dimensions 5 μm wide and 0.7 μm high and with optical losses 1.36 dB/cm at 810 nm and 2.01 dB/cm at 980 nm, were presented in [7].

In this paper, we presented optical waveguides with optical losses lower or comparable with our previously presented results [9,17,29] and also comparable with results presented by Choi et al. [30]. Bamiedakis et al. presented an optical waveguide for high-speed applications with very low optical losses operating at 850 nm [14], but our applied R2P fabrication process allowed low cost and mass production. Our presented waveguides can also be used for the wider wavelength range from visible to infrared spectrum.

5. Conclusions

The paper has presented the properties of the planar optical waveguides fabricated by the roll-to-plate imprinting. PDMS stamps were fabricated from Sylgard 184 elastomer using a nickel master mold and the process was optimized for fabrication optical channels with dimension 50 × 50 μm and 250 μm pitch. These dimensions of the waveguides were chosen due to multimode fiber optics communications standards and increasing interest in new high-speed on-board optical interconnects used in data centers and in super high-speed computers.

The development of the fabrication process included optimizing the distance between the glass plate with the polymer layer and the R2P cylinder, the speed of the plate moving under the R2P cylinder and the UV curing dose applied during the R2P process and also post-processing UV curing. We also optimized the process of the deposition core layer by doctor blading technology and the sharp razor blade proved to be the most appropriate for our applied polymers.

The presented waveguides will allow the realization of a flexible optical waveguiding structure suitable for various applications. Due to their low optical losses, they would be useful for operations not only in the infrared spectrum at 850 and 1300 nm, which are the regions of major interest in optical interconnections but also in the visible spectral region which can be applied for optical visible communications or optical sensors. The paper has demonstrated that R2P NIL has great potential for the implementation of optical polymer waveguides not only for optical interconnection applications, and that the fabrication procedure allowed easy and mass production of the optical waveguide devices.

The next research will be focused to fabricate single-mode waveguides using R2P technology. We estimate dimensions for the single-mode waveguides from 0.5 μm up to 4 μm depends on the refractive index contrast for the applied materials and used transmitted wavelengths. The critical technology process for this purpose is to fabricate a high-quality PDMS stamp. Therefore, it is necessary to fabricate a high-quality master. We will use the laser beam writing process for the fabrication of the master stamp with the dimension of the single-mode waveguides and we also believe that it will be possible to then fabricated single-mode waveguides using R2P technology.

Author Contributions: V.P. and P.K. conceived and designed the experiments; V.C. and S.K. performed the experiments; V.C., V.P., and M.N. have measured properties of the channel waveguides; M.N. measured optical losses; V.P. performed transmission spectra measurement and dark modes spectroscopy; V.P. has analysed the data and wrote the paper; M.M. review the paper. All authors have read and agreed to the published version of the manuscript.

Funding: This work was supported by the CTU grant no. SGS20/175/OHK3/3T/13 as well as by the Centre of Advanced Applied Natural Sciences, Reg. No. CZ.02.1.01/0.0/0.0/16_019/0000778, supported by the Operational Program Research, Development and Education, co-financed by the European Structural and Investment Funds and the state budget of the Czech Republic. P.K., S.K. and M.M. acknowledge funding from the NIL2 industry project.

Institutional Review Board Statement: The funders had no role in the design of the study; in the collection, analyses, or interpretation of data; in the writing of the manuscript, or in the decision to publish the results.

Informed Consent Statement: Not applicable.

Data Availability Statement: The data presented in this study are available on request from the corresponding author.

Acknowledgments: The authors would like to thank for technical support Karsten Winters from BASF Lampertheim GmbH.

Conflicts of Interest: The authors declare no conflict of interest.

References

1. Dong, P.; Chen, Y.-K.; Duan, G.-H.; Neilson, D.T. Silicon photonic devices and integrated circuits. *Nanophotonics* **2014**, *3*, 215–228. [CrossRef]
2. Wong, W.; Liu, K.; Chan, K.; Pun, E. Polymer devices for photonic applications. *J. Cryst. Growth* **2006**, *288*, 100–104. [CrossRef]
3. Rahlves, M.; Rezem, M.; Boroz, K.; Schlangen, S.; Reithmeier, E.; Roth, B. Flexible, fast, and low-cost production process for polymer based diffractive optics. *Opt. Express* **2015**, *23*, 3614–3622. [CrossRef] [PubMed]
4. Eldada, L.; Shacklette, L. Advances in polymer integrated optics. *IEEE J. Sel. Top. Quantum Electron.* **2000**, *6*, 54–68. [CrossRef]
5. Ma, H.; Jen, A.K.Y.; Dalton, L.R. Polymer-based optical waveguides: Materials, processing, and devices. *Adv. Mater.* **2002**, *14*, 1339–1365. [CrossRef]
6. Prajzler, V.; Hyps, P.; Mastera, R.; Nekvindova, P. Properties of Siloxane Based Optical Waveguides Deposited on Transparent Paper and Foil. *Radioengineering* **2016**, *25*, 230–235. [CrossRef]
7. Beche, B.; Pelletier, N.; Gaviot, E.; Zyss, J. Single-mode TE_{00}-TM_{00} optical waveguides on SU-8 polymer. *Opt. Commun.* **2004**, *230*, 91–94. [CrossRef]
8. Elmogi, A.; Bosman, E.; Missinne, J.; Steenberge Van, G. Comparison of epoxy- and siloxane-based single-mode optical waveguides defined by direct-write lithography. *Opt. Mater.* **2016**, *52*, 26–31. [CrossRef]
9. Prajzler, V.; Neruda, M.; Jasek, P.; Nekvindova, P. The properties of free-standing epoxy polymer multi-mode optical waveguides. *Microsyst. Technol.* **2019**, *25*, 257–264. [CrossRef]
10. Ibrahim, M.H.; Kassim, N.M.; Mohammad, A.B.; Lee, S.-Y.; Chin, M.-K. Single mode optical waveguides based on photodefinable benzocyclobutene (BCB 4024–40) polymer. *Microw. Opt. Technol. Lett.* **2006**, *49*, 479–481. [CrossRef]
11. Chemoptics Inc. Available online: http://www.chemoptics.co.kr/eng/main/main.php (accessed on 12 March 2021).
12. Prajzler, V.; Jasek, P.; Nekvindova, P. Inorganic–organic hybrid polymer optical planar waveguides for micro-opto-electro-mechanical systems (MOEMS). *Microsyst. Technol.* **2019**, *25*, 2249–2258. [CrossRef]
13. Buestrich, R.; Kahlenberg, F.; Popall, M.; Dannberg, P.; Muller-Fiedler, R.; Rosch, O. ORMOCER (R) s for optical interconnection technology. *J. Sol-Gel Sci. Technol.* **2001**, *20*, 181–186. [CrossRef]
14. Bamiedakis, N.; Beals, J.; Penty, R.V.; White, I.H.; DeGroot, J.V.; Clapp, T.V. Cost-Effective Multimode Polymer Waveguides for High-Speed On-Board Optical Interconnects. *IEEE J. Quantum Electron.* **2009**, *45*, 415–424. [CrossRef]
15. Bosman, E.; Van Steenberge, G.; Christiaens, W.; Hendrickx, N.; Vanfleteren, J.; Van Daele, P. Active optical links embedded in flexible substrates. In Proceedings of the 58th Electronic Components and Technology Conference, Lake Buena Vista, FL, USA, 27–30 May 2008; pp. 1150–1157.
16. Cai, Z.; Qiu, W.; Shao, G.; Wang, W. A new fabrication method for all-PDMS waveguides. *Sens. Actuators A Phys.* **2013**, *204*, 44–47. [CrossRef]
17. Prajzler, V.; Neruda, M.; Nekvindová, P. Flexible multimode polydimethyl-diphenylsiloxane optical planar waveguides. *J. Mater. Sci. Mater. Electron.* **2018**, *29*, 5878–5884. [CrossRef]
18. Kagami, M.; Kawasaki, A.; Ito, H. A polymer optical waveguide with out-of-plane branching mirrors for surface-normal optical interconnections. *J. Lightwave Technol.* **2001**, *19*, 1949–1955. [CrossRef]
19. Klein, S.; Barsella, A.; Leblond, H.; Bulou, H.; Fort, A.; Andraud, C.; Lemercier, G.; Mulatier, J.; Dorkenoo, K. One-step waveguide and optical circuit writing in photopolymerizable materials processed by two-photon absorption. *Appl. Phys. Lett.* **2005**, *86*, 211118. [CrossRef]
20. Wong, W.H.; Zhou, J.; Pun, E.Y.B. Low-loss polymeric optical waveguides using electron-beam direct writing. *Appl. Phys. Lett.* **2001**, *78*, 2110–2112. [CrossRef]
21. Wolfer, T.; Bollgruen, P.; Mager, D.; Overmeyer, L.; Korvink, J.G. Flexographic and Inkjet Printing of Polymer Optical Waveguides for Fully Integrated Sensor Systems. *Procedia Technol.* **2014**, *15*, 521–529. [CrossRef]
22. Choi, C.-G.; Han, S.-P.; Kim, B.C.; Ahn, S.-H.; Jeong, M.-Y. Fabrication of large-core 1×16 optical power splitters in polymers using hot-embossing process. *IEEE Photon. Technol. Lett.* **2003**, *15*, 825–827. [CrossRef]
23. Fan, R.; Hooker, R. Tapered polymer single-mode waveguides for mode transformation. *J. Lightwave Technol.* **1999**, *17*, 466–474. [CrossRef]
24. Kobayashi, J.; Yagi, S.; Hatakeyama, Y.; Kawakami, N. Low Loss Polymer Optical Waveguide Replicated from Flexible Film Stamp Made of Polymeric Material. *Jpn. J. Appl. Phys.* **2013**, *52*, 072501. [CrossRef]
25. Bruck, R.; Muellner, P.; Kataeva, N.; Koeck, A.; Trassl, S.; Rinnerbauer, V.; Schmidegg, K.; Hainberger, R. Flexible thin-film polymer waveguides fabricated in an industrial roll-to-roll process. *Appl. Opt.* **2013**, *52*, 4510–4514. [CrossRef]
26. Shneidman, A.V.; Becker, K.P.; Lukas, M.A.; Torgerson, N.; Wang, C.; Reshef, O.; Burek, M.J.; Paul, K.; McLellan, J.; Loncar, M. All-polymer integrated optical resonators by roll-to-roll nanoimprint lithography. *ACS Photonics* **2018**, *5*, 1839–1845. [CrossRef]
27. Shao, J.; Chen, X.; Li, X.; Tian, H.; Wang, C.; Lu, B. Nanoimprint lithography for the manufacturing of flexible electronics. *Sci. China Ser. E Technol. Sci.* **2019**, *62*, 175–198. [CrossRef]

28. Yde, L.; Lindvold, L.; Stensborg, J.; Voglhuber, T.; Außerhuber, H.; Wögerer, S.; Fischinger, T.; Mühlberger, M.; Hackl, W. Roll-to-Plate UV-nanoimprinting for micro and nano-optics. In Proceedings of the 15th International Conference on Nanoimprint & Nanoprint Technology, Braga, Portugal, 26–28 September 2016.
29. Prajzler, V.; Neruda, M.; Květoň, M. Flexible multimode optical elastomer waveguides. *J. Mater. Sci. Mater. Electron.* **2019**, *30*, 16983–16990. [CrossRef]
30. Lin, L.; Liu, Y.; Choi, J.; Wang, L.; Haas, D.; Magera, J.; Chen, R.T. Flexible optical waveguide film fabrications and optoelectronic devices integration for fully embedded board-level optical inter-connects. *J. Lightwave Technol.* **2004**, *22*, 2168–2175.

 *nanomaterials*

Article

Fabrication of High Aspect Ratio Micro-Structures with Superhydrophobic and Oleophobic Properties by Using Large-Area Roll-to-Plate Nanoimprint Lithography

Nithi Atthi [1,*], Marc Dielen [2], Witsaroot Sripumkhai [1], Pattaraluck Pattamang [1], Rattanawan Meananeatra [1], Pawasuth Saengdee [1], Oraphan Thongsook [1], Norabadee Ranron [1], Krynnaras Pankong [1], Warinrampai Uahchinkul [1], Jakrapong Supadech [1], Nipapan Klunngien [1], Wutthinan Jeamsaksiri [1], Pim Veldhuizen [2] and Jan Matthijs ter Meulen [2]

[1] Thai Microelectronics Center (TMEC), National Electronics and Computer Technology Center (NECTEC), Chachoengsao 24000, Thailand; witsaroot.sripumkhai@nectec.or.th (W.S.); pattaraluck.pattamang@nectec.or.th (P.P.); rattanawan.meananeatra@nectec.or.th (R.M.); pawasuth.saengdee@nectec.or.th (P.S.); oraphan.thongsook@nectec.or.th (O.T.); norabadee.ran@nectec.or.th (N.R.); krynnaras.pan@ncr.nstda.or.th (K.P.); warinrampai.uah@ncr.nstda.or.th (W.U.); jakrapong.supadech@nectec.or.th (J.S.); nipapan.klunngien@nectec.or.th (N.K.); wutthinan.jeamsaksiri@nectec.or.th (W.J.)

[2] Morphotonics B.V., De Run 4281, 5503 LM Veldhoven, The Netherlands; marc.dielen@morphotonics.com (M.D.); pim.veldhuizen@morphotonics.com (P.V.); jan.matthijs.ter.meulen@morphotonics.com (J.M.t.M.)

* Correspondence: nithi.atthi@nectec.or.th

Citation: Atthi, N.; Dielen, M.; Sripumkhai, W.; Pattamang, P.; Meananeatra, R.; Saengdee, P.; Thongsook, O.; Ranron, N.; Pankong, K.; Uahchinkul, W.; et al. Fabrication of High Aspect Ratio Micro-Structures with Superhydrophobic and Oleophobic Properties by Using Large-Area Roll-to-Plate Nanoimprint Lithography. *Nanomaterials* **2021**, *11*, 339. https://doi.org/10.3390/nano11020339

Academic Editor: Yang-Tse Cheng

Received: 5 January 2021

Accepted: 23 January 2021

Published: 29 January 2021

Publisher's Note: MDPI stays neutral with regard to jurisdictional claims in published maps and institutional affiliations.

Abstract: Bio-inspired surfaces with superamphiphobic properties are well known as effective candidates for antifouling technology. However, the limitation of large-area mastering, patterning and pattern collapsing upon physical contact are the bottleneck for practical utilization in marine and medical applications. In this study, a roll-to-plate nanoimprint lithography (R2P NIL) process using Morphotonics' automated Portis NIL600 tool was used to replicate high aspect ratio (5.0) micro-structures via reusable intermediate flexible stamps that were fabricated from silicon master molds. Two types of Morphotonics' in-house UV-curable resins were used to replicate a micro-pillar (PIL) and circular rings with eight stripe supporters (C-RESS) micro-structure onto polycarbonate (PC) and polyethylene terephthalate (PET) foil substrates. The pattern quality and surface wettability was compared to a conventional polydimethylsiloxane (PDMS) soft lithography process. It was found that the heights of the R2P NIL replicated PIL and C-RESS patterns deviated less than 6% and 5% from the pattern design, respectively. Moreover, the surface wettability of the imprinted PIL and C-RESS patterns was found to be superhydro- and oleophobic and hydro- and oleophobic, respectively, with good robustness for the C-RESS micro-structure. Therefore, the R2P NIL process is expected to be a promising method to fabricate robust C-RESS micro-structures for large-scale anti-biofouling application.

Keywords: high aspect ratio micro-structure; roll-to-plate nanoimprint lithography; superhydrophobic; oleophobic; biomimetic surface; large-area patterning

1. Introduction

Biofouling from colonization of various organisms, pathogens and inorganic macro-molecules (mostly proteins) is the unwanted accumulation of biological and inorganic matters on wetted surfaces [1]. The contamination of surfaces, such as marine infrastructure, medical devices and other engineering components, has been a global issue with significant impact on the environment, health risks and economics [1,2]. Biofouling by numerous pathogens such as viruses, bacteria, fungi and other infectious agents causes a spread of infectious diseases in public space which potentially leads to thousands of annual deaths worldwide [1]. Being submerged in seawater, surfaces of a ship hull are exposed to thousands of species of fouling organisms such as barnacles, mussels and seaweed. Issues include marine corrosion and increased ship hull drag, which result in reduced speed, increased fuel consumption and emissions of greenhouse gases (CO_2, NO_x, and SO_2) [3]. On top of that, it costs the global economy USD 150 billion yearly due to the environmental and significant economic impacts [2,3].

For antibacterial and antifouling solutions in medical environments, such as a surgery room and ward area, expensive materials such as Teflon and paint with nanoparticles (NPs) are commonly used. However, the processes to fabricate silver (Ag), titanium dioxide (TiO_2) and zinc oxide (ZnO) NPs as well as the manufacturing and the spray coating of the paint with non-bonded NPs can be toxic if not taking careful precautionary measures [4–6]. In 2008, the use of biocide-based paint that contains toxic substances such as Tributhyltin (TBT) and lead (Pb) was banned because of its severe effect on living organisms in the ocean [7]. Subsequently, non-biocide-release approaches, such as water jet cleaning and ultrasonic wave methods, were proposed. However, these technologies are very harmful for the operator working in underwater environment. Moreover, the ultrasonic waves can interfere the communication of marine mammals [8,9]. Therefore, alternative methods that are inexpensive, robust, easy to apply, scalable, and environmentally friendly are required. Super dewetting or antifouling surfaces have shown excellent resistance to stains, bacteria, proteins, and various marine organisms due to the absence of effective adhesion points on these surfaces.

The super dewetting surfaces exhibit good anti-biofouling and self-cleaning properties, which have the potential to address the concerns of the above-mentioned traditional biocide-based antifouling methods. There are three types of antifouling surfaces: superhydrophobic surfaces, underwater superoleophobic surfaces, and slippery liquid-infused porous surfaces (SLIPS). Among these, superhydrophobic surfaces, with a contact angle of water (surface tension, γ_{lv} = 72.1 mN/m) greater than 150° and a sliding angle smaller than 10°, are one of the promising technologies for antifouling purpose [8–12]. Such superhydrophobic surfaces are often inspired by nature. For instance, it is reported that the surfaces of various natural animals and plants, such as lotus leaves, red rose petals, butterfly wings, mosquito eyes, Salvinia leaves, water strider legs, gecko feet, and shark skin, show exceptional wetting behavior with superhydrophobicity [13–18].

Basically, surface wettability is categorized by the contact angle of the liquid droplet [19,20]. Firstly, if the water contact angle (WCA) is smaller than 90°, a surface is considered hydrophilic. Secondly, if the WCA is greater than 90°, it is considered hydrophobic. Lastly, if the WCA is greater than 150°, it is considered superhydrophobic [21]. Typically, the contact angle of liquid droplets on a flat surface can be explained by Young's model as shown in Figure 1a. However, a surface is never completely smooth and generally exhibits a surface roughness. The effect of surface roughness on the apparent contact angle of liquid droplets can be explained by the Wenzel and Cassie-Baxter models as shown in Figure 1b,c, respectively [22].

The Wenzel model is used to describe the wetting behavior of a rough surface by calculating the apparent contact angle of the textured surface ($\theta_w{}^*$) based on its surface roughness (r), surface tension between solid–gas (γ_{sv}), solid–liquid (γ_{sl}), and liquid–gas (γ_{lv}) interfaces, and an intrinsic contact angle of the liquid droplet on a flat surface of the same material (θ), as shown in Equation (1) [22–24]. In the Wenzel state, the liquid wets the surface and completely fills all voids on the rough surface. The contact angle in the Wenzel state is defined by:

$$\cos\theta_W* = \frac{r(\gamma_{sv} - \gamma_{sl})}{\gamma_{lv}} = r\cos\theta \tag{1}$$

in which the surface roughness factor r is defined by the ratio of the actual surface area of the rough surface (A_h) to the projected area (A_f), as shown in Equation (2):

$$r = 1 + \frac{A_h}{A_f} \tag{2}$$

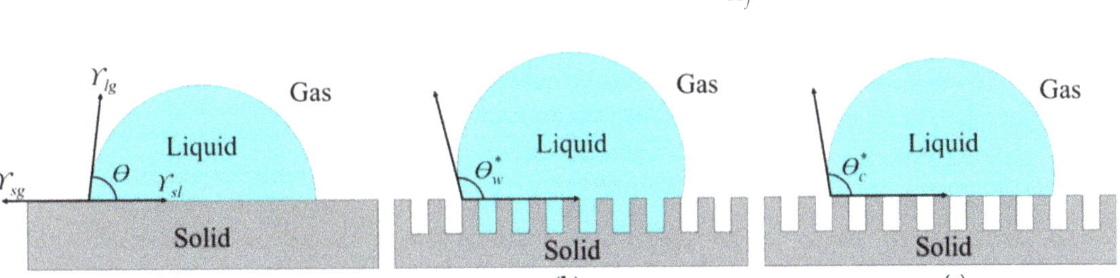

Figure 1. A liquid droplet resting on a solid surface and surrounded by a gas forms a characteristic contact angle θ. (**a**) Flat surface; (**b**) Rough surface at which the liquid is in intimate contact with the solid asperities, so that the droplet is in the Wenzel state; (**c**) Rough surface at which the liquid rests on top of the asperities, so that the droplet is in the Cassie-Baxter state [20,21]. Note that θ_w^* is the apparent contact angle of the textured surface based on the Wenzel model, θ_c^* is the apparent contact angle of the textured surface based on the Cassie-Baxter model, and γ_{sv}, γ_{sl}, γ_{lv} is the surface tension between solid-gas, solid-liquid, and liquid-gas interfaces, respectively.

The Cassie-Baxter model describes that the ultimate liquid-repellent nature of the rough surface is caused by microscopic air pockets filled in the space between the rough micro- or nano-structures and the liquid droplet. The air pockets then create a combination of air–liquid–solid interfaces [13,21,25]. If ϕ_s is the fraction of the solid in contact with the liquid, the Cassie-Baxter equation can be expressed as Equation (3).

$$\cos\theta_C^* = -1 + \phi_s(1 + \cos\theta) \tag{3}$$

To maximize the WCA in Wenzel's model, Young's contact angle (θ), as determined on a flat surface, has to be minimized and the surface roughness (r) has to be maximized [26]. There are two methods to construct a superhydrophobic surface. The first one is to reduce the surface energy of the material, and the other is to increase the surface roughness. In the past decades, many research groups designed different hierarchical micro-/nano-structures on low surface energy material to produce a biomimetic superhydrophobic surface. The value of r of the patterned surface has to be maximized by reducing the pattern width (a) and pattern spacing (b) down to the nanometer scale, and by increasing the pattern height (h) [27–29]. This is shown schematically in Figure 2a–c, respectively, in which the surface roughness r of an example micro-/nano-structure is increased by increasing the packing factor (P = a/b) and the aspect ratio (A.R. = h/a).

Note that here the pattern width is equal to the pattern spacing ($a_1 = b_1$, $a_2 = b_2$, $a_3 = b_3$) but where $a_1 < (a_2 = a_3)$ and ($h_1 = h_2$) < h_3. The highest packing factor can be achieved with the smallest surface structures (i.e., nano-structures). These are typically made in a conventional lithography process. In making such silicon (Si) master molds, the resolution of the lithographic exposure tool and performance of the deep reactive ion etching (DRIE) tool limits a maximum value of the packing factor and a maximum value of the aspect ratio of the hierarchical micro-/nano-structures.

As structured substrate material, Polydimethylsiloxane (PDMS) is commonly known for its low surface energy (γ_{sv} = 12–16 mJ/m^2), low Young's modulus ($\sim$2.0 MPa), good thermal and oxidative stability, non-toxicity, good biocompatibility, and low cost. PDMS also provides a conformal contact and it is released easily from a Si master mold [30,31]. Moreover, PDMS has a great degree of flexibility during the patterning process due to its relatively high toughness and high elongation at break (>150%) [32,33]. Based on all of these benefits combined, PDMS becomes one of the best material choices that is widely used to fabricate surfaces with superhydrophobic properties using a soft lithography process [32,34,35]. A flat PDMS surface only shows hydrophobic properties with a WCA of 107–110° [36]. Therefore, the fabrication of a rough surface with engineered hierarchical micro-/nano-structures having controlled geometries by the soft lithography process is needed to make PDMS become superhydrophobic [33,34]. Regarding our previous work, PDMS surfaces patterned with a conventional square-like pillar pattern arranged in a square array can demonstrate superhydrophobicity (WCA > 150°) when r and A.R. are greater than 2.75 and 3.0, respectively [27]. To obtain a superoleophobic and superamphiphobic surface, the A.R. of the micro-/nano-structures must even be greater than 5.0.

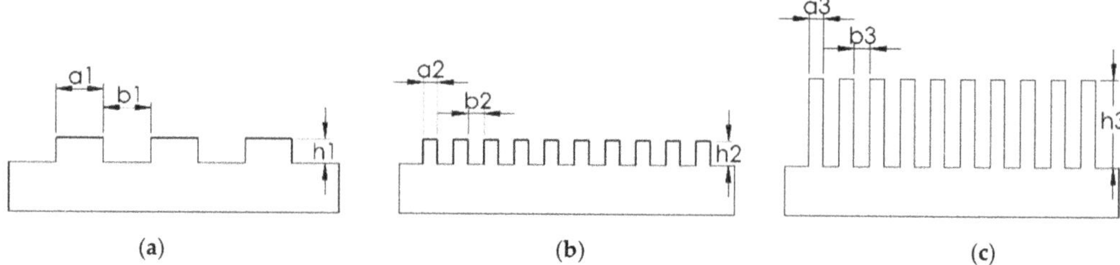

Figure 2. Schematic diagram of an example micro-/nano-structure having a different surface roughness (r) due to a different packing factor (P = a/b) and aspect ratio (A.R. = h/a), where a is the pattern width, b is the pattern spacing, and h is the pattern height. (**a**) Low P and low A.R., (**b**) high P and low A.R., and (**c**) high P and high A.R. pattern.

Typically, there are two ways to create a nature-inspired superhydrophobic surface with hierarchical micro-/nano-structured rough surface on a substrate with low surface energy: replication of a biological surface by a casting (i.e., soft lithography) process or an engineered rigid master template in a "top-down approach" photolithographic method. The soft lithography process is widely studied because it is easy and accurate to replicate the micro-/nano-patterns with various A.R. on the relatively soft PDMS substrate. However, the flexibility of PDMS (σ: 5.0 MPa, ε: 116% @RT) limits its feasibility in the processability of high A.R. structures and increasing pattern density, hydrophobicity, and mechanical strength. The challenges of using PDMS also include deformation, merging, and collapsing of structures, which result in decreased hydrophobicity [35].

Therefore, there are three main approaches to improve the robustness and strength of these superhydrophobic surfaces. Firstly, a well-designed robust micro-/nano-structure is required to prevent the PDMS pattern from collapsing and maintain its superhydrophobicity [36]. Secondly, utilization of PDMS-based composite materials to improve its mechanical properties [37,38]. Note that the surface energy of the PDMS-based composite materials is often higher, which adversely affects the superhydrophobic properties. Thirdly, development of new materials, besides PDMS, with low surface energy and high mechanical strength.

To realize the full potential of antifouling surfaces, engineered hierarchical micro-/nano-structures with superhydrophobic and superoleophobic properties need to be produced over a large area in a cost-efficient manner for practical applications [39]. This continuous and efficient fabrication of superhydrophobic surfaces on large areas is challenging and few processes have been adopted by industrial concerns. This is due to the fact that most of the reported methods require multiple complex processing steps, and have low throughput, substrate limitations, or high production cost [40–44]. Among many conventional micro- and nano-patterning techniques that have been developed in the past few decades, the roll-to-plate ultraviolet (UV) nanoimprint lithography (R2P NIL) and the roll-to-roll UV nanoimprint lithography (R2R NIL) technologies are some of the promising solutions due to the lower cost, higher throughput, larger patterning area, and higher resolution beyond the limitations set by light diffraction or beam scattering that are encountered in other traditional micro-/nano-fabrication techniques [44].

One of the technical challenges of the UV NIL process is the scaling of the master. As the patterns have to be transferred from a Si master mold (or other master material) in order to fabricate a flexible stamp that can be reused, the fabrication of large-area Si molds tends to be difficult as the feature sizes go down to lower ranges of the nanometer scale. Therefore, often a scaling up of a small area on a (expensive) master mold containing the micro- or nano-structures needs to be performed. Additionally, the material selection for the UV NIL process is also crucial in overcoming critical issues such as the well-known stamp sticking problem, polymerization shrinkage, and thermal and hygroscopic expansion as well as prolonged lifetime of the flexible stamp [44]. Moreover, the replication of micro-/nano-structures becomes challenging when the A.R. is greater than 2.0 because of stamp releasing problems due to increased contact area.

Since most superhydrophobic surfaces can be easily contaminated by different types of fat- and oil-based liquids that have much lower surface tension, such as decane (γ_{lv} = 23.8 mN/m) and octane (γ_{lv} = 21.6 mN/m) [45], the superhydrophobic applications are limited and challenged in several situations [25]. Instead of using superhydrophobic surfaces, superoleophobic surfaces (organic liquids repellent) and superamphiphobic surfaces (water and oil repellent) are more likely to be used for water-/oil-proof properties in polluted water or greasy environment [25]. Therefore, this study extends to an investigation of the superoleophobicity of the micro-structures.

In this research, the fabrication of high A.R. micro-structures with superhydrophobicity and oleophobicity using R2P NIL process was investigated and demonstrated. The effect of applying a low surface energy resin to replicate micro-structures having A.R. 5.0 onto both polycarbonate (PC) and polyethylene terephthalate (PET) foil substrates on the pattern qualities and pattern fidelity was studied. The effects of the imprinted hierarchical micro-structures on the hydrophobicity, oleophobicity and robustness were investigated as well. Furthermore, the performance of the R2P NIL process in fabricating the water- and oil-repellent properties and robustness of the hierarchical micro-structures has been compared to conventional PDMS micro-structures fabricated by a soft lithography process.

2. Materials and Methods

2.1. Pattern Design

In this research, two different micro-structures, including a square-like pillar (PIL) pattern and circular rings with eight stripe supporters (C-RESS) pattern were designed as shown schematically in Figure 3a,b, respectively. These patterns were arranged in a square array. Si master molds of both the PIL and C-RESS micro-structure were fabricated by a conventional photolithography and deep reactive ion etching (DRIE) process [34,36]. In this experiment, the resolution of the lithographic exposure tool called Nikon stepper, model NSR2005i8A (Miyagi Nikon Precision Co., Ltd., Miyaki, Japan) is 0.5 μm. Therefore, the pattern was designed with a size (a) of 0.5 μm, a pattern spacing (b) of 0.5 μm, and a pattern height (h) of 2.5 μm. Hence, the A.R. of the PIL and C-RESS patterns were 5.0.

2.2. Silicon Mold Fabrication

First, the surface of the 6 inch p-Si (100) wafers was cleaned by standard cleaning (SC-1) process using the automatic wet bench (AWB) machine (STEAG Energy Services GmbH, Essen, North Rhine-Westphalia, Germany) to remove unwanted contaminants [46]. Then, a 2.0 μm-thick silicon dioxide (SiO_2) layer was deposited onto the Si wafers by means of a low-pressure chemical vapor deposition (LPCVD) process using Samco SVG TMX 2604 (Samco-ucp Ltd., Ruggell Liechtenstein). The O_2 and H_2 gas flow rate is 5.0 and 8.5 standard liter per minutes (SLPM), respectively. The oxidation temperature is 1050 °C with a duration time of 13 h. Layers of 1.0 μm-thick Sumitomo PFI-34A photoresist (PR) (Sumitomo chemical advanced technologies, Phoenix, AZ, USA) with either the PIL or C-RESS pattern were fabricated by conventional photolithography process using the

Nikon stepper, model NSR2005i8A (Miyagi Nikon Precision Co., Ltd., Miyaki, Japan). The Hg lamp power is 395 mW/cm^2, exposure time is 920 ms, focus distance is −0.4 µm. After being developed with the Sumitomo SD-W developer (Sumitomo chemical advanced technologies, Phoenix, AZ, USA) for 80 s, the PR patterns were transferred to the SiO$_2$ layer by reactive ion etching (RIE) process (CF$_4$/CHF$_3$/Ar: 10/50/100 sccm, P: 150 mTorr, RF: 800 W/10 min) using the RIE system from Applied Materials, model AMAT P5000 Mark II (Applied Materials, Inc., Santa Clara, CA, USA). Later, the patterns with a height of 2.5 µm were etched into the Si by DRIE from Oxford Instruments, model Plasmalab ICP100 (Oxford Instruments Plasma Technology, Bristol, UK) using Bosch process for 10 cycles, respectively, as shown in Figure 4a. One etching cycle includes a deposition step (C$_4$F$_8$/SF$_6$: 300/5 sccm, ICP/RIE: 2000/0 W, 6 s) and an etching step (C$_4$F$_8$/SF$_6$: 1/200 sccm, ICP/RIE: 2000/15 W, 5 s). After stripping the PR by using oxygen plasma ashing (Ramco RAM-250, Ramco Equipment Corporation, Hillside, NJ, USA) process (RF power 1000 W, O$_2$: 900 sccm, gas pressure: 900 mTorr at 80 °C for 1 h), the Si wafers were immersed in piranha acid solution (70 wt% H$_2$SO$_4$: 30 wt% H$_2$O$_2$, 4:1 weight ratio) from JT Baker® (Avantor, Phillipsburg, NJ, USA) at 120 °C for 10 min. Then the samples were rinsed with deionized water (DIW) for 15 min and blown dry with pure nitrogen gas. Finally, the remained SiO$_2$ hard mask was etched by buffer oxide etching (BOE) process using HF (49% vol): NH$_4$F (40% vol) of ratio 1:7 from JT Baker® (Avantor, Phillipsburg, NJ, USA) at 25 °C for 30 min with an etching rate of 900 angstrom per minute.

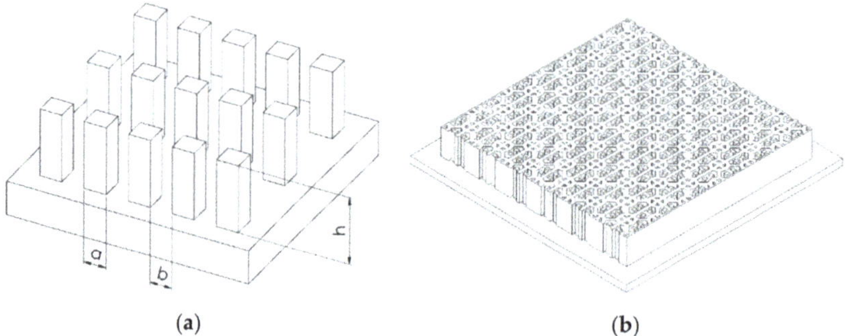

(a)   (b)

Figure 3. Schematic diagram of the fabricated micro-structures (**a**) square-like pillar (PIL) and (**b**) circular rings with eight stripe supporters (C-RESS) pattern.

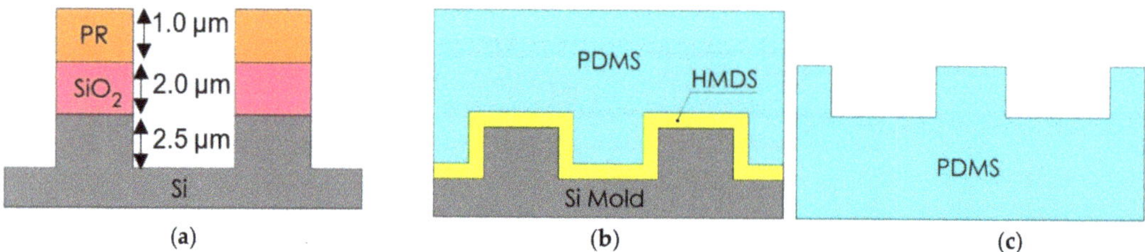

Figure 4. Cross-sectional schematic diagram (not to scale) of (**a**) PR/SiO$_2$/Si patterning by conventional photolithography and plasma etching process (**b**) HMDS priming and PDMS casting by soft lithography process, and (**c**) peeled off PDMS sample from the Si master mold.

2.3. Soft Lithography Process for PDMS Replication

First, the hydrophilic Si master molds were primed in hexamethyldisilaxane (HMDS) vapor (JT Baker® 9362-09, Avantor, Phillipsburg, NJ, USA) in a desiccator at 25 °C for 48 h to improve their demolding properties as shown in Figure 4b. This prevents getting mold damage from hard PDMS residues in a poor demolding process. The HMDS molecules form a self-assembled monolayer (SAM) onto the Si surface of which the non-polar methyl groups repel water droplets inducing a slight hydrophobicity onto the Si surface (WCA: $60°$–$80°$) with correspondingly good PDMS mold release properties [34,36]. Subsequently, the PDMS mixture (Sylgard 184, curing agent: 10 wt%, Sigma-Aldrich, St. Louis, MO, USA) was poured onto the Si master molds containing either the PIL or C-RESS pattern [34,36]. After curing in a convection oven at 75 °C for 120 min, the PDMS-PIL and PDMS-C-RESS patterns were released from their Si master mold by peeling them by hand, resulting in samples as schematically shown in Figure 4c. A flat PDMS (PDMS-FLT) surface was also fabricated from a non-patterned Si surface as a control sample.

2.4. Roll-to-Plate NIL Process for Large-Area Patterning

The R2P NIL process was carried out onto Morphotonics' automated Portis NIL600 tool (Morphotonics, Veldhoven, The Netherlands) as shown in Figure 5a [47]. Before imprinting, the as-received patterned Si master molds were primed using an in-house releasing agent treatment (Morphotonics, Veldhoven, The Netherlands) with a surface energy of 10–15 mN/m). Subsequently, an in-house transparent flexible stamp is copied from the primed Si master molds in a UV-NIL imprint process. This flexible stamp, containing the inverse PIL or C-RESS micro-structure from the Si master molds and having anti-stick properties, was mounted around the rollers in the Portis NIL600 imprint tool (Morphotonics, Veldhoven, The Netherlands) and an in-house developed UV-curable resin (Morphotonics, Veldhoven, The Netherlands) was dispensed (manually) onto a PC or PET foil substrate which was placed onto a carrier as shown in Figure 5b. Then, the coated substrate was moved to the nip of the imprint rollers and the imprint was performed as follows: as the substrate moves through the nip, the negative structure on the flexible stamp is transferred onto the substrate by laminating the stamp onto the substrate with the UV-curable resin by exerting a uniform line pressure with the imprint roller at a fixed speed. Subsequently, the resin is cured through the transparent stamp by a 365 nm UV-light emitting diode (UV-LED) bar (the UV intensity was set at 0.4 W/cm^2) after which the stamp is delaminated from the imprinted substrate by the delamination roller. The flexible stamp is then transported back to the start position and can be reused for a next imprint cycle. This can be repeated for more than 1000 times with a single flexible stamp. A schematic of the R2P NIL process is depicted in Figure 6.

(a)   (b)

Figure 5. (**a**) Morphotonics' automated Portis NIL600 tool (Morphotonics, Veldhoven, The Netherlands) and (**b**) set-up of a flexible stamp used for imprinting a substrate coated with UV-curable resin (Morphotonics, Veldhoven, The Netherlands) [47].

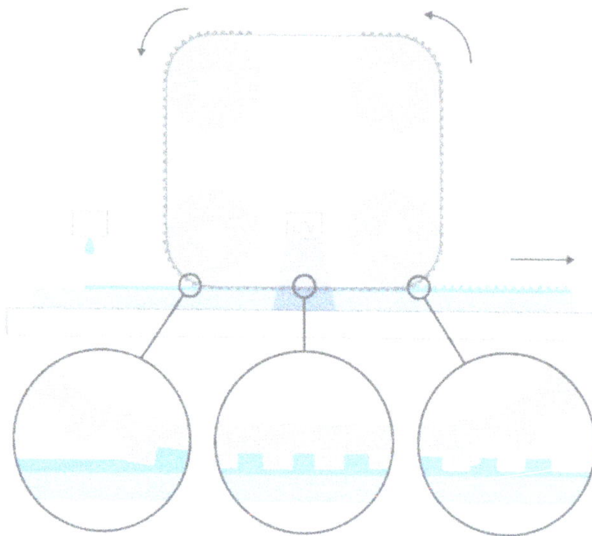

Figure 6. Schematic diagram of Morphotonics' roll-to-plate ultraviolet nanoimprint lithography (R2P NIL) process (Morphotonics, Veldhoven, The Netherlands) [47].

For all samples the imprint speed was 0.2 m/min, though it is possible to imprint faster in order to increase throughput (up to 2 m/min was tested for the PIL micro-structure). The gap between imprint roller and substrate was set to −0.75 mm. Both the PIL and C-RESS micro-structures could be imprinted with a broad range of line pressures (2.5–15.0 N/cm), thereby accurately controlling the residual layer thickness. For the C-RESS pattern often the adhesion of either the stamp or the imprint was failing due to the large forces during delamination (due to large surface area), despite using specifically developed in-house resins with low surface free energy (SFE) and good adhesion to both PC and PET. In particular, the two in-house resins that have been used, are: MM1078 (Morphotonics, Veldhoven, The Netherlands), hereafter resin A (viscosity: 149 mPa·s, surface energy: ~15 mN/m, Young's modulus: 59 MPa) and MM2138C4 (Morphotonics, Veldhoven, The Netherlands), hereafter resin B (viscosity: ~35.3 mPa·s, surface energy: ~10 mN/m, Young's modulus: ~170 MPa).

2.5. Surface Characterizations

After imprinting, the R2P NIL fabricated flexible stamps and samples were characterized by means of a 3D laser scanning confocal microscope (Keyence, VK-X1000 series, Keyence Corporation, Osaka, Japan) in order to inspect the replicated pattern quality and overall defects. Subsequently, the pattern shape and surface topology of both the PDMS and R2P NIL samples were characterized by a field-emission scanning electron microscope (FE-SEM, Hitachi S-4700, Hitachi group, Tokyo, Japan). All samples were coated with 2.0 nm-thick platinum to improve the conductivity of the PDMS and resin patterns.

Droplets of 5.0 μL DIW and ethylene glycol (EG) (Sigma-Aldrich, St. Louis, MO, USA) were dropped onto the patterned surfaces on five different locations of each sample. The measurement was conducted to obtain an average water (γ_{lv} = 72.1 mN/m) [48] contact angle (WCA) and ethylene glycol (γ_{lv} = 47.3 mN/m) [48] contact angle (EGCA) by a contact angle goniometer (Ramé-hart instrument Co., Model-400, Succasunna, NJ, USA). The tilt angle of the substrate was 0°. Test conditions are stable in the cleanroom class 1000 with well-controlled environment temperature at 22 ± 1 °C and relative humidity at 45 ± 5%. The contact angle is estimated using the sessile drop technique by measuring the angle between the tangent lines along solid-liquid interface and liquid-vapor interface of the liquid contour.

The advancing contact angle (θ_{adv}) was measured by adding a volume of liquid to the first drop dynamically for five times (drop volume has increased from 5.0 to 25.0 µL) to the maximum volume of 25.0 µL permitted without increasing the three-phase line. The receding contact angle (θ_{rec}) was measured by removing a volume of liquid from the drop for 4 times. The drop volume has then decreased from 25.0 to 5.0 µL. When the receding contact angle is subtracted from the advancing contact angle, the result is called the contact angle hysteresis (CAH). The CAH characterizes surface topology and can help quantify contamination, surface chemical heterogeneity, and the effect of surface treatments, surfactants and other solutes [49].

The SFE was measured by a mobile surface analyzer, MSA (KRÜSS GmbH, Hamburg, Germany) and extracted by using Owens, Wendt, Rabel and Kaelble (OWRK) method. In the OWRK model, at least two liquids such as DIW (γ_{lv} = 72.1 mN/m) and diiodomethane (DI) from Sigma-Aldrich, St. Louis, MO, USA (γ_{lv} = 50.8 mN/m) with known dispersive (γ_{sv}^{D}) and polar parts (γ_{lv}^{P}) of the surface tensions are needed to compute the solid SFE [50]. The total SFE (γ_{sv}) of the solid is the sum of the two parts as described in Equation (4). The results enable well-founded statements about wettability by aqueous or organic liquids.

$$\gamma_{sl} = \gamma_{sv} + \gamma_{lv} - 2\left(\sqrt{\gamma_{sv}^{D}\gamma_{lv}^{D}} + \sqrt{\gamma_{sv}^{P}\gamma_{lv}^{P}} \right) \tag{4}$$

γ_{sv}^{D} and γ_{lv}^{D} are the dispersive components and γ_{sv}^{P} and γ_{lv}^{P} are the polar components of the solid and liquid surface energies, respectively.

Lastly, surface durability was tested by scratching the PDMS and R2P NIL fabricated micro-structures using a glass slide with an applied compressive force while moving the glass slide back and forth 30 times.

3. Results and Discussion

3.1. PDMS Pattern Qualities and Surface Properties

In case of the PDMS pattern design, the dimensions of the PIL and C-RESS micro-structures on the Si master molds are (a = 2.0 µm, b = 2.0 µm, h = 2.5 µm) and (a = 1.0 µm, b = 1.5 µm, h = 2.5 µm), respectively. Therefore, the A.R. of these PIL and C-RESS Si micro-structures are 1.25 and 2.50, respectively, which are lower than of the PIL and C-RESS micro-structures on the Si master molds used for the R2P NIL replication processes. Nevertheless, the pattern heights are the same so that a comparison between the replication quality and pattern fidelity of the PDMS soft lithography and R2P NIL fabrication processes can be done to this extent. Note that the feature size was limited to 1.0 µm because here, the Si mold patterning was done using a contact mask aligner (EVG model 620, EV Group Europe & Asia/Pacific GmbH, St. Florian am Inn, Austria). Furthermore, here the Si master molds contain the inverse PIL (i.e., holes) and C-RESS micro-structures because the PDMS soft lithography process consists of only one casting step as opposed to the R2P NIL process used in this study. Figure 7a,b show the SEM images of the PDMS-PIL and PDMS-C-RESS replicated micro-structures before scratch test, respectively. It can be seen that these PDMS patterns were successfully released from the HMDS primed Si master molds. Both PDMS micro-structures were well replicated from the Si molds without defects like pattern clumping, collapsing or breaking off. It is observed that the PDMS patterns have very smooth surfaces without sidewall scallops from the Si mold features, as well as less steep vertical sidewalls. This could possibly have been caused by insufficient curing or a (thermal) reflow effect of the PDMS patterns.

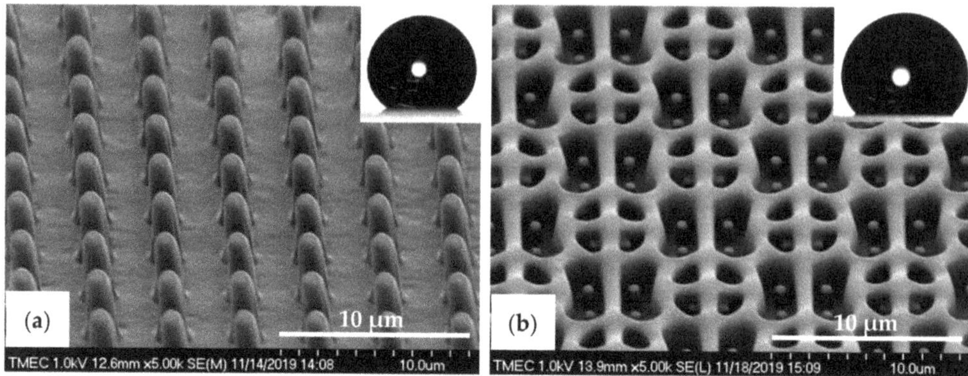

Figure 7. Tilted-view SEM images of the PDMS micro-structures fabricated by soft lithography process before scratch test. (**a**) PDMS-PIL pattern (designed pattern size: diameter: 2.0 µm, height: 2.5 µm) and (**b**) PDMS-C-RESS pattern (designed pattern size: diameter: 1.0 µm, height: 2.5 µm). The insets show water droplet contact angle images on those PDMS surfaces.

The cross-sectional SEM images in Figure 8a,b show that the pattern width (a) of the PDMS-PIL (a: 1.9 ± 0.02 µm) and PDMS-C-RESS patterns (a: 0.9 ± 0.10 µm) was well controlled resulting in a −5.0% and −10.0% deviation (i.e., shrinkage) from the pattern design, respectively. Note that the vertical sidewalls are less steep so that the width at the top of the pattern is smaller than at the bottom. The pattern heights (h) were measured at 4.8 ± 0.06 µm for the PDMS-PIL and 2.3 ± 0.07 µm for the PDMS-C-RESS pattern resulting in a 92.0% elongation and an 8.0% shrinkage compared to the pattern design, respectively. These numbers result in an A.R. of 2.53 and 2.56 for the PDMS-PIL and PDMS-C-RESS micro-structures, respectively. The differences of the PDMS pattern heights could be explained by a number of factors. Firstly, the pattern depths on the Si master molds could differ due to the micro-loading and the aspect ratio dependent etching (ARDE) effects, which cause the Si etch rate to be dependent on the density and A.R. of the micro-structures, respectively [51–53]. Secondly, it could be that the flexible PDMS patterns are stretched during peeling from the Si mold (possibly the case for the PDMS-PIL samples) or have shrunk during curing (possibly the case for the PDMS-C-RESS samples).

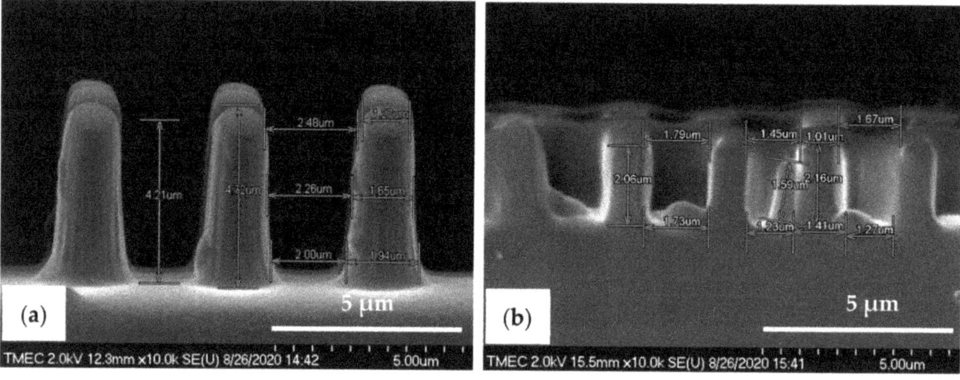

Figure 8. Cross-sectional view SEM images with corresponding measured dimensions of the PDMS micro-structures fabricated by soft lithography process before scratch test. (**a**) PDMS-PIL pattern (designed pattern size: diameter: 2.0 µm, height: 2.5 µm) and (**b**) PDMS-C-RESS pattern (designed pattern size: diameter: 1.0 µm, height: 2.5 µm).

In order to investigate the surface wettability behavior, the WCA of the PDMS-PIL and PDMS-C-RESS micro-structures before the scratch test were measured as 145 ± 3.2°

and 130 $\pm$ 2.4°, respectively. Both patterned PDMS samples did not show any superhydrophobic properties with the WCA greater than 150°. This is due to the still relatively low A.R. of ~2.5 of the designed and subsequently replicated micro-structures. In addition, CAH of the PDMS-PIL and PDMS-C-RESS micro-structures are 1.5° and 10.7°, respectively, showing a better stability of the surface wettability for the PIL pattern.

Next, the robustness of the PDMS replicated patterns was studied by means of a scratch test. After the scratch test, the PDMS-PIL pattern was mostly found collapsed due to the applied external forces, as shown in Figure 9a. The pattern collapsing was caused by the poor mechanical strength of the PDMS. Furthermore, pattern mating and clumping were also observed after the scratch tests. These defects are due to the fact that the Van Der Waals force between the adjacent patterns is larger than pulling force or recovery force. Moreover, the electrostatic discharge (ESD) induced during the scratch test can generate an adhesion force between the pillars. In contrast, the PDMS-C-RESS pattern did not collapse after the scratch tests and did not exhibit other defects or damage, as shown in Figure 9b. This is due to the robust design of the C-RESS micro-structure pattern.

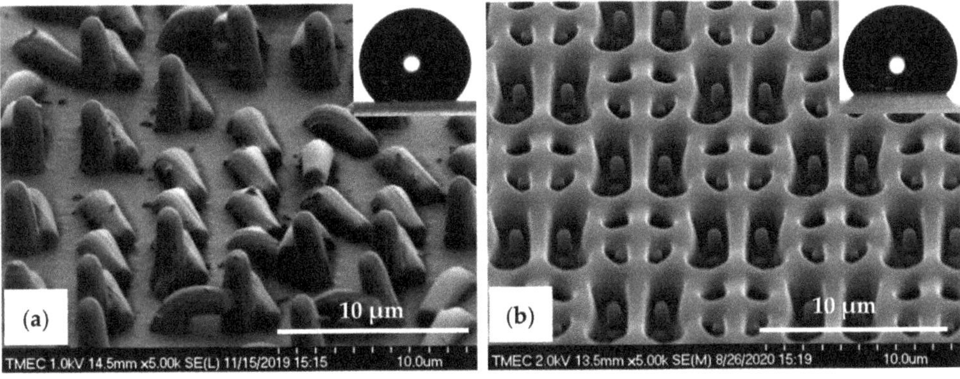

Figure 9. Tilted-view SEM images of the PDMS micro-structures fabricated by soft lithography process after scratch test. (a) PDMS-PIL pattern (designed pattern size: diameter: 2.0 µm, height: 2.5 µm) and (b) PDMS-C-RESS pattern (designed pattern size: diameter: 1.0 µm, height: 2.5 µm). The insets show water droplet contact angle images on those PDMS surfaces.

After scratch tests, the WCA of the PDMS-PIL sample was decreased to 135.9 $\pm$ 9.5° due to the induced defects, while the WCA of the PDMS-C-RESS pattern stayed quite the same at 130.0 $\pm$ 4.9°, respectively. It was also found that the standard deviation (S.D.) of the WCA of the PDMS-PIL pattern has increased after scratch tests. Moreover, the CAH of the PDMS-PIL pattern has increased significantly to 21.4°. This is due to the variation of the surface roughness across the PDMS-PIL surface due to the pattern collapsing, mating, and clumping. In contrast, the S.D. and CAH (11.5°) of the PDMS-C-RESS patterns did not really increase because the C-RESS micro-structure is robust enough to maintain the pattern shape and surface roughness after the scratch test. Based on the results, the C-RESS pattern is considered attractive for antifouling applications because of its robustness and already quite high hydrophobicity. However, the pattern width and spacing should be further reduced and the pattern height should be further increased to reach an A.R. of 5.0, which is minimally needed in order to obtain superhydrophobicity as well as superoleophobic properties.

3.2. R2P NIL Pattern Qualities and Surface Properties

The R2P NIL imprints have been made using another Si master mold. The laser confocal microscopy image of the pillar (PIL) micro-structure on the Si mold in Figure 10a shows that the DRIE process at TMEC was well-controlled resulting in a high-quality pattern with no defects or contamination. This confocal microscope gives a general quality indication. Due to the steep angles the confocal microscope cannot determine the structure dimensions (and therefore not replication fidelity) accurately. SEM characterization and step profilometer measurement confirmed that the etch depth of the Si micro-structures was found uniform across a six-inch Si wafer with only 2.0% deviation from the pattern design (data not shown). A precise and uniform flexible stamp (intermediate resin mold) containing the inverse polarity of the PIL pattern (i.e., holes) was well fabricated from the in-house releasing agent treated Si master mold, as shown in Figure 10b. Using the R2P NIL process, the pillar micro-structure from the Si mold (A.R. of 5.0) was well replicated onto both PC and PET foil substrates with both resin A (PIL-A) and resin B (PIL-B), as shown in Figure 10c,d. According to the laser confocal microscopy images, no pattern clumping, collapsing or breaking off of the micro-pillars was found on the Si master mold and the imprinted samples PIL-A and PIL-B, as shown in Figure 10a,c,d, respectively. Moreover, no air bubble defects were present on the samples after imprinting. These results are further corroborated by SEM analysis of the samples, which gives a more detailed view of the replicated pattern quality and allows for an accurate determination of the replication fidelity. This is important for a good evaluation of the antifouling behavior of the imprinted micro-structures. The SEM images of the pillar micro-structures imprinted with resin A (PIL-A) and resin B (PIL-B) before scratch test are shown in Figure 11a,b, respectively. It can be seen that, with the aid of Morphotonics' in-house Si mold releasing agent and due to the low SFE of the used resins, the micro-pillars were well replicated from the Si master mold and subsequently from the flexible stamp within the R2P NIL process. Again, no pattern clumping, collapsing or breaking down was found as well as no air bubble defects. Furthermore, in Figure 12a,b, the cross-sectional SEM images show that the pattern width (a) of the PIL-A (a: 458.3 ± 20 nm) and PIL-B patterns (a: 586.0 ± 37 nm) was different resulting in a −8.3% and +17.2% deviation from the original design, respectively, despite having experienced the same R2P NIL fabrication process.

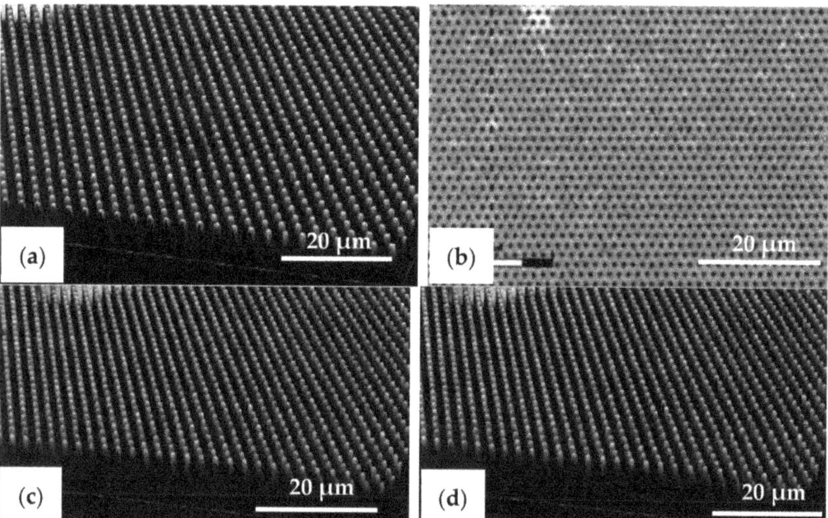

Figure 10. Laser confocal microscopy images of the pillar micro-structure (designed pattern size: diameter: 0.5 μm, height: 2.5 μm) on the (**a**) Si master mold before in-house releasing agent treatment, (**b**) R2P NIL flexible stamp (holes, top-view), (**c**) imprint on PC foil with resin A (PIL-A), and (**d**) imprint on PC foil with resin B (PIL-B).

Figure 11. Tilted-view SEM images of the pillar micro-structure (designed pattern size: diameter: 0.5 μm, height: 2.5 μm) replicated by R2P NIL process (**a**) imprint on PC foil with resin A (PIL-A) and (**b**) imprint on PC foil with resin B (PIL-B).

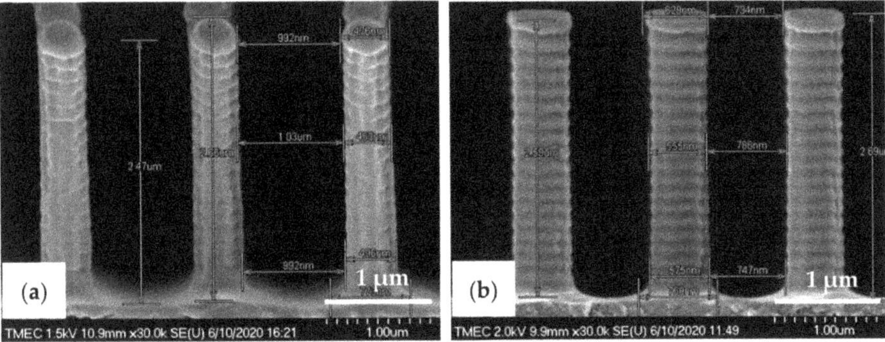

Figure 12. Cross-sectional view SEM images with corresponding measured dimensions of the pillar micro-structure (designed pattern size: diameter: 0.5 μm, height: 2.5 μm) replicated by R2P NIL process (**a**) imprint on PC foil with resin A (PIL-A) and (**b**) imprint on PC foil with resin B (PIL-B).

The almost vertical sidewalls and rough surfaces with sidewall scallops from the Si master mold were also observed on the PIL-A and PIL-B patterns, while they were absent on the PDMS samples. This underlines the high patterning resolution of the R2P NIL process. The pattern height (h) was measured at 2.57 ± 0.13 μm and 2.64 ± 0.07 μm, respectively. Calculation of the imprinted pattern height deviation from the original design gives a 2.8 and 5.6% elongation resulting in an A.R. of 5.6 and 4.5 for the PIL-A and PIL-B micro-structures, respectively. These differences are likely attributed to the discrepancy between the original design and the actual Si mold pattern heights and to the different resin properties which result in different polymerization shrinkage values upon UV curing and different flexibilities during delamination, amongst others. Here, due to the large surface area the forces during delamination are such high that the cured micro-pillars could have been stretched out more than they have shrunk.

Besides replicating the micro-pillar pattern, also fabrication of the more robust C-RESS micro-structure with the R2P NIL process was investigated. Without Morphotonics' in-housing Si mold releasing agent treatment (i.e., with the HMDS vapor primed Si wafer) the C-RESS structure was not successfully replicated. The laser confocal microscopy images in Figure 13 show that the patterns are completely delaminated/broken off from the substrate, as can been seen from the comparison with the Si master mold pattern. This is caused by the high forces during delamination of the flexible stamp from the cured patterns, which is due to the large surface area of the C-RESS micro-structure and the relatively high SFE of the HMDS primed Si master mold and, subsequently, of the flexible stamp, despite the use of the low SFE resins.

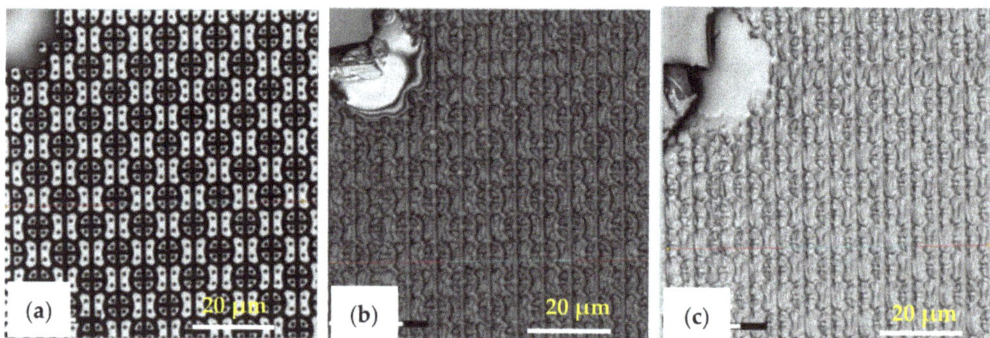

Figure 13. Top-view laser confocal microscopy images of the C-RESS micro-structure (pattern design: size: 1.5 μm, space: 0.5 μm, height: 2.5 μm) on the (**a**) Si master mold before in-house releasing agent treatment (i.e., with HMDS vapor priming), (**b**) imprint on PET foil with resin A (C-RESS-A), and (**c**) imprint on PET foil with resin B (C-RESS-B). Note that the artefact seen in the top left corner is part of the design on the Si master mold and is therefore copied onto the samples.

With the aid of Morphotonics' in-house Si mold releasing agent, however, the replication of the C-RESS pattern was significantly improved though still challenging and several defects were observed. Firstly, as shown in Figure 14, it is observed on the imprinted C-RESS-A and C-RESS-B samples that many of the small holes are filled instead of well-replicated. This could be explained by the incomplete filling of the holes in the Si master mold during fabrication of the flexible stamp or by the breaking off of the pillars on the flexible stamp during delamination. Secondly, it can be seen that sometimes also other parts of the C-RESS pattern can be broken off, again indicating the high forces that are present during delamination. Nevertheless, these results show that the choice of Si mold releasing agent in conjunction with the used resins and flexible stamp materials and the imprint settings is key in achieving a good replication of the high A.R. C-RESS micro-structure.

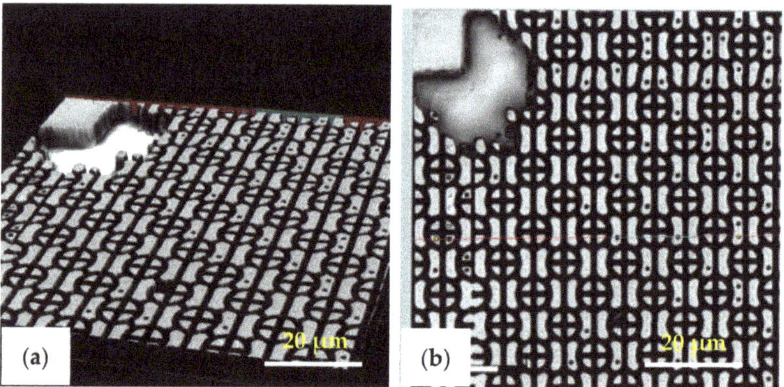

Figure 14. Laser confocal microscopy images of the C-RESS micro-structure (pattern design: size: 1.5 μm, space: 0.5 μm, height: 2.5 μm) on the (**a**) imprint on PC foil with resin A (C-RESS-A), and (**b**) imprint on PC foil with resin B (C-RESS-B).

Here, the intermediate flexible stamp (not shown) used was made from the Si master mold after the in-house releasing agent treatment. Note that the artefact seen in the top left corner is part of the design on the Si master mold (see Figure 13a) and is, therefore, copied onto the samples. Detailed SEM imaging of the C-RESS-A and the C-RESS-B micro-structures before the scratch test further corroborate the abovementioned results and are shown in Figure 15a–c, respectively. Although the first two SEM images seem to presume that the replications were perfect and without defects, they are very local recordings of the patterned area. Moreover, a third SEM image shows indeed that many of the filled holes seem to be due to the breaking off of the pillars on the flexible stamp during delamination. Furthermore, no pattern mating, clumping or collapsing of the robust C-RESS micro-structure was found on the imprints as well as no air bubble defects. The cross-sectional SEM images in Figure 16a,b show that the pattern width (a) of the C-RESS-A (a: 1.52 ± 0.11 µm) and PIL-B patterns (a: 1.53 ± 0.04 µm) was practically the same and only deviating 1.3% and 2.0% from the original design, respectively.

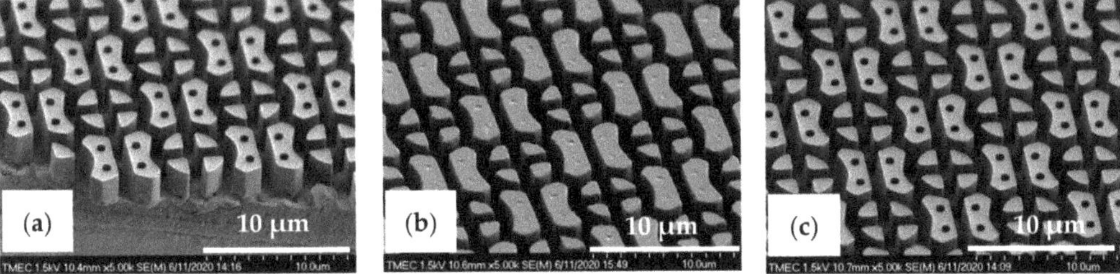

Figure 15. Tilted-view SEM images of C-RESS micro-structure (pattern design: size: 1.5 µm, space: 0.5 µm, height: 2.5 µm) replicated by R2P NIL process (**a**) imprint on PC foil with resin A (C-RESS-A), and (**b,c**) imprints on PC foil with resin B (C-RESS-B).

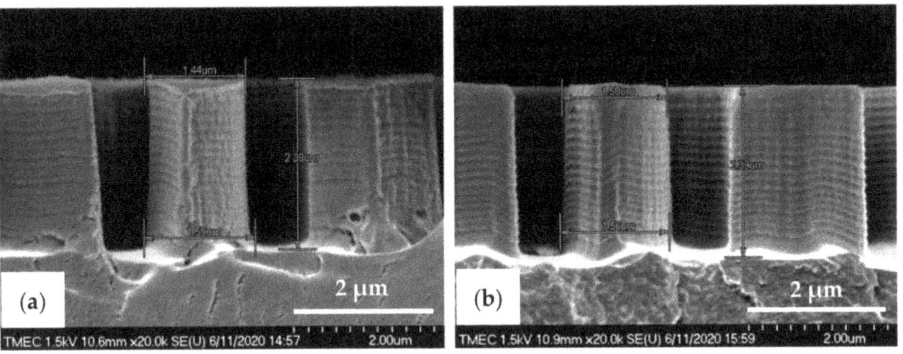

Figure 16. Cross-sectional view SEM images with corresponding measured dimensions of the C-RESS micro-structure (pattern design: size: 1.5 µm, space: 0.5 µm, height: 2.5 µm) replicated by R2P NIL process (**a**) imprint on PC foil with resin A (C-RESS-A) and (**b**) imprint on PC foil with resin B (C-RESS-B).

Similar to the pillar micro-structure the C-RESS-A and C-RESS-B replicated patterns also exhibit the rough surfaces with sidewall scallops due to the Si master mold fabrication (DRIE process), again showing the high patterning resolution of the R2P NIL process. The pattern height (h) was measured at 2.38 ± 0.04 µm and 2.39 ± 0.06 µm, respectively. This results in a pattern height shrinkage of 4.8% and 4.4% compared to the original design. Since the pattern fidelity values for both the C-RESS-A and C-RESS-B samples are so close to each other, it may be concluded that the two resins exhibit similar shrinkage upon UV curing (for this particular micro-structure). It is likely the challenging design of the C-RESS pattern that limits a flawless replication.

The difference between the replicated PIL and the C-RESS pattern heights (but also part of the observed deviations from the pattern design) can be explained by a different pattern depth on the Si master molds due to the earlier mentioned micro-loading and ARDE effects, which cause the Si etch rate to be dependent on the density and A.R. of the micro-structures [51–53]. Taking this into consideration, it may be that the actual pattern height fidelity of the C-RESS micro-structure replications are possibly even better and that the A.R. could be close to 5.0.

In order to examine the surface wetting behavior, each pattern was fabricated multiple times to investigate the reproducibility of the R2P NIL process by measuring sample-to-sample deviations. Therefore, the micro-structure replicated with resin A was labeled with sample identification numbers (sample ID) as PIL-A-1, PIL-A-2, C-RESS-A-1, and C-RESS-A-2 and the micro-structure replicated with resin B as PIL-B-1, PIL-B-2, C-RESS-B-1, and C-RESS-B-2. The WCA of the PIL-A, PIL-B, C-RESS-A, and C-RESS-B samples before the scratch test was measured as shown in Table 1 and the EGCA was measured as shown in Table 2. In addition, the advancing and receding WCA and EGCA are presented in Tables 1 and 2 as well as the respective CAH.

Table 1. Measured water contact angle (WCA) and its contact angle hysteresis (CAH) of the PIL and C-RESS micro-structures replicated with resin A and resin B on PC foil. Note that for each sample the WCA was measured on five different locations across the sample.

Scheme	WCA [°]	WCA_{adv} [°]	WCA_{rec} [°]	CAH [°]
PIL-A-1	152.9 ± 0.8	151.4 ± 2.2	150.2 ± 2.2	1.1 ± 0.6
PIL-A-2	153.8 ± 1.3	151.8 ± 1.4	148.5 ± 1.2	1.7 ± 1.3
PIL-B-1	153.3 ± 1.8	152.5 ± 0.5	151.5 ± 0.4	1.0 ± 0.3
PIL-B-2	151.6 ± 2.5	150.7 ± 0.6	149.5 ± 1.0	1.3 ± 0.8
C-RESS-A-1	147.1 ± 3.6	141.8 ± 2.4	130.9 ± 3.6	10.4 ± 2.5
C-RESS-A-2	144.6 ± 2.9	137.9 ± 1.4	131.5 ± 1.4	6.9 ± 1.7
C-RESS-B-1	147.1 ± 2.9	141.8 ± 3.3	130.0 ± 3.8	11.8 ± 2.0
C-RESS-B-2	144.6 ± 2.1	137.9 ± 1.8	129.2 ± 1.5	8.8 ± 2.7

Table 2. Measured ethylene glycol contact angle (EGCA) and its contact angle hysteresis (CAH) of the PIL and C-RESS micro-structures replicated with resin A and resin B on PC foil. Note that for each sample the EGCA was measured on five different locations across the sample.

Sample ID	EGCA [°]	$EGCA_{adv}$ [°]	$EGCA_{rec}$ [°]	CAH [°]
PIL-A-1	148.8 ± 1.3	150.7 ± 2.1	147.3 ± 1.2	3.4 ± 1.4
PIL-A-2	149.3 ± 1.2	149.7 ± 0.9	147.9 ± 0.9	1.7 ± 0.5
PIL-B-1	151.0 ± 1.7	150.2 ± 2.0	148.9 ± 0.6	1.4 ± 1.3
PIL-B-2	150.8 ± 1.7	150.9 ± 1.0	149.3 ± 0.7	1.6 ± 0.8
C-RESS-A-1	145.4 ± 3.9	144.1 ± 4.1	128.8 ± 5.9	15.3 ± 8.5
C-RESS-A-2	144.5 ± 2.7	137.4 ± 2.6	125.4 ± 2.3	11.8 ± 2.3
C-RESS-B-1	145.0 ± 1.3	140.4 ± 1.7	127.4 ± 2.0	12.9 ± 1.2
C-RESS-B-2	142.2 ± 3.8	139.2 ± 3.1	125.1 ± 2.4	14.1 ± 1.4

Furthermore, the WCA, the EGCA and their corresponding droplet shapes on the different micro-structures before scratch test are shown in Figure 17a,b, respectively. From a first glance, it can be concluded that there is no significant difference of the surface wettability between the patterns replicated with resin A and resin B. This shows that the SFE of the cured imprint resins does not necessarily has to be very low, because the A.R. of the micro-structure patterns themselves (~5.0) already induce a large portion of the surface (super) hydrophobicity. In order to make a strong conclusion about this, the micro-structures should be replicated using higher SFE resins as well. These higher SFE will have the difficulty that, most probably, the imprint will fail due to stamp sticking problems during delamination. Figure 18a,b show the CAH of the water and the ethylene glycol (EG) droplets on the various replicated micro-structure surfaces. The water CAH of the PIL and the C-RESS micro-structures are in the range of $1.0 \pm 0.3°$ to $1.7 \pm 1.3°$ and $6.9 \pm 1.7°$ to $11.8 \pm 2.0°$, respectively. The EG CAH of the PIL and the C-RESS micro-structures are in the range of $1.4 \pm 1.3°$ to $3.4 \pm 1.4°$ and $11.8 \pm 2.3°$ to $15.3 \pm 8.5°$.

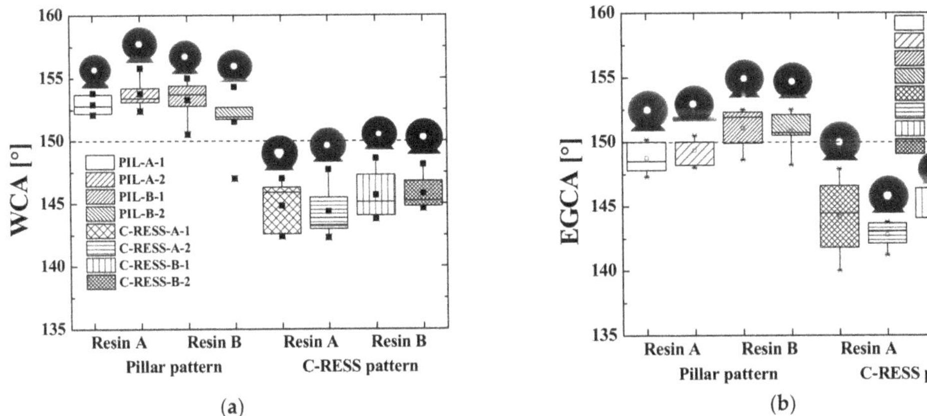

Figure 17. Effects of micro-structure pattern and resin type on the contact angle of (**a**) water (WCA) and (**b**) ethylene glycol (EGCA) of the imprinted samples.

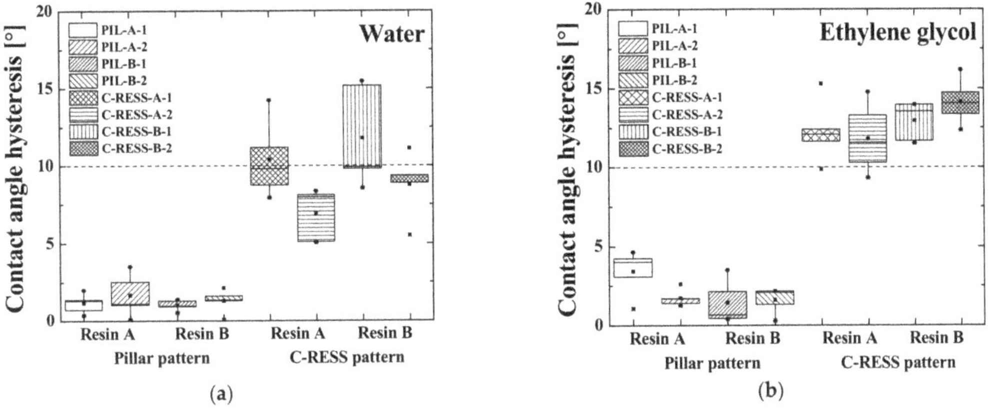

Figure 18. Effects of micro-structure pattern and resin type on the contact angle hysteresis (CAH) of (**a**) water droplets and (**b**) ethylene glycol droplets dispensed on the imprinted samples.

From this, it can be concluded that the surface wettability of the PIL micro-structure is more stable than that of the C-RESS micro-structure because of the lower value of CAH. This can be explained by the better imprint quality and fidelity of the PIL micro-structure than compared to the C-RESS micro-structure. In addition, the PIL-A and the PIL-B micro-structures show superhydrophobicity and superoleophobicity because the WCA and the EGCA were greater than 150° and the value of the water CAH and EG CAH was lower than 10°. In contrast, the C-RESS-A and the C-RESS-B micro-structures show only hydrophobicity and oleophobicity, although the WCA and EGCA values are not that much lower than of the PIL-A and PIL-B samples. This is likely due to the fact that the C-RESS pattern was replicated less well by the R2P NIL process and with more artefacts (such as the filled holes) compared to the PIL pattern. Nevertheless, if the pattern fidelity of the C-RESS micro-structure can be improved, the surface wettability may become superhydrophobic and superoleophobic.

Furthermore, the hydrophobicity (and oleophobicity) of both the PIL and C-RESS patterns replicated by the R2P NIL process is higher than that of the PDMS samples fabricated by the (conventional) soft lithography process. This is mostly due to the higher A.R. of the patterns on the R2P NIL samples (~5.0) compared to those on the PDMS samples (~2.5). In addition, there could also be differences in the surface wettability due to the different properties of the used materials. However, since the SFE of the used resins and of PDMS are quite similar, these differences are expected to be very minor. Nevertheless, it could be that the better replication quality of the R2P NIL samples (as opposed to the smoothed PDMS patterns) results in the very high WCA and EGCA values in combination with the higher A.R. Moreover, the R2P NIL patterns showed lower water CAH and EG CAH compared to the PDMS samples. This means that the R2P NIL process has a higher patterning resolution and higher pattern fidelity with better uniformity across the sample surfaces compared to the conventional soft lithography process.

3.3. Effects of Micro-Structure Pattern and Resin Types on the Surface Energy

Table 3 shows the WCA and diiodomethane contact angle (DICA) of the PIL-A, PIL-B, C-RESS-A, and C-RESS-B imprinted patterns. Note that the data were obtained using the mobile surface analyzer (MSA) system (KRÜSS GmbH, Hamburg, Germany), which is a different contact angle measurement tool than what has been used in the previous section. Therefore, these contact angle values are slightly different.

Table 3. Calculated surface free energy (SFE) values of the PIL and C-RESS micro-structures replicated with two different resins on PC foil, as measured by using water and diiodomethane.

Sample ID	WCA [°]	DICA [°]	γ_{sv}^{D} [mN/m]	γ_{sv}^{P} [mN/m]	γ_{sv} [mN/m]
PIL-A-1	145.3 ± 5.9	132.7 ± 2.9	1.31 ± 0.3	0.02 ± 0.1	1.33 ± 0.4
PIL-A-2	151.6 ± 2.1	129.7 ± 2.9	1.65 ± 0.4	0.05 ± 0.1	1.70 ± 0.4
PIL-B-1	142.1 ± 6.2	112.6 ± 5.5	4.81 ± 1.4	0.13 ± 0.3	4.94 ± 1.7
PIL-B-2	151.2 ± 1.5	114.7 ± 6.3	4.30 ± 1.5	0.53 ± 0.4	4.83 ± 1.9
C-RESS-A-1	146.4 ± 3.1	143.0 ± 1.3	0.90 ± 1.1	0.46 ± 1.3	1.38 ± 2.5
C-RESS-A-2	153.8 ± 1.5	140.0 ± 2.8	0.70 ± 0.2	0	0.70 ± 0.2
C-RESS-B-1	141.3 ± 3.9	134.0 ± 3.9	1.19 ± 3.4	0.17 ± 0.2	1.35 ± 0.6
C-RESS-B-2	139.8 ± 5.8	129.8 ± 3.4	1.65 ± 0.4	0.50 ± 0.7	2.15 ± 1.1

The WCA values of all samples were higher than their DICA values, as shown in Figure 19a. This is an effect of the lower value of the water surface tension (γ_{lv} = 72.1 mN/m) compared to that of diiodomethane from Sigma-Aldrich, St. Louis, MO, USA (γ_{lv} = 50.8 mN/m). The values of the WCA and DICA were used to calculate the SFE (γ_{sv}) of the imprinted samples using the OWRK model. It was found that the γ_{sv} of the PIL-A and PIL-B patterns before the scratch test was in the range of 1.33 ± 0.4 to 1.70 ± 0.4 mN/m and 4.83 ± 1.9 to 4.94 ± 1.7 mN/m, respectively, and that of the C-RESS-A and C-RESS-B patterns in the range of 0.70 ± 0.2 to 1.38 ± 2.5 mN/m and 1.35 ± 0.6 to 2.15 ± 1.1 mN/m, respectively. Figure 19b shows that the γ_{sv} of the micro-structures imprinted with resin A is lower than that of the micro-structures imprinted with resin B, even though the SFE of a flat sample of resin A (γ_{sv}~15 mN/m) is higher than that of a flat sample of resin B (γ_{sv}~10 mN/m). This difference is especially pronounced for the PIL pattern. This effect is not very well understood yet, since it contrasts to the results in the previous section. However, it is possible that there could be variations in replication quality and pattern fidelity between the samples. Moreover, since resin B is still quite experimental, it could be that its properties change over time or are inhomogeneous across multiple samples, thereby causing this discrepancy. More research should be performed in order to fully understand this.

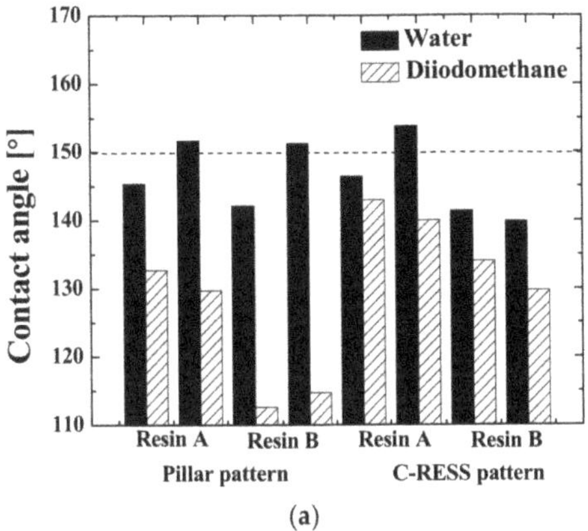
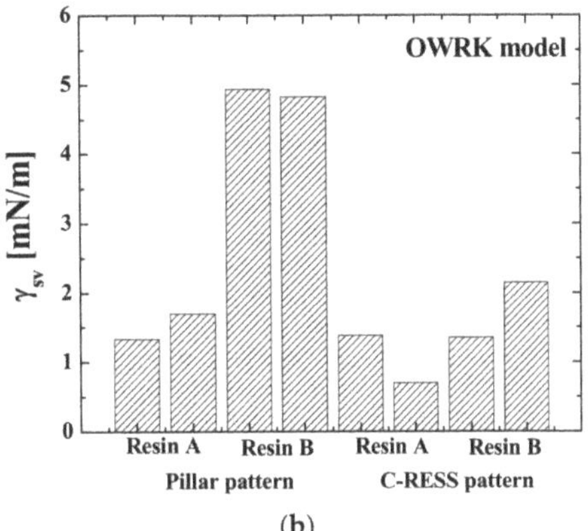

Figure 19. Effects of micro-structure pattern and resin type on the surface properties (**a**) water and diiodomethane contact angle and (**b**) surface free energy of the imprinted micro-structures.

3.4. Robustness of the R2P NIL Imprinted Micro-Structures after Scratch Test

After performing scratch tests, SEM analysis was carried out again and the WCA and EGCA were remeasured in order to investigate the robustness of the R2P NIL replicated micro-structures. The PIL-A and PIL-B micro-structures were mostly found collapsed, twisted or even broken off by the applied external forces, as is shown in Figure 20. Besides pillar collapse and the removal of pillars also pattern mating and clumping were observed. This is due to the fact that the Van Der Waals force between the pillars is larger than the pulling force or recovery force [54]. Moreover, the ESD induced during the scratch test can generate an adhesion force between the pillars. The collapsing or breaking off of the micro-pillars was caused by their poor (inherent) mechanical strength and the low Young's modulus of the used resins. Although the latter is higher for the used R2P NIL resins than compared to PDMS in the reference soft lithography fabrication process, it is still insufficient for use as a robust anti-biofouling coating. To this end, the C-RESS micro-structure pattern was designed. After the scratch tests, both the C-RESS-A and C-RESS-B patterns remained intact, as shown in Figure 21.

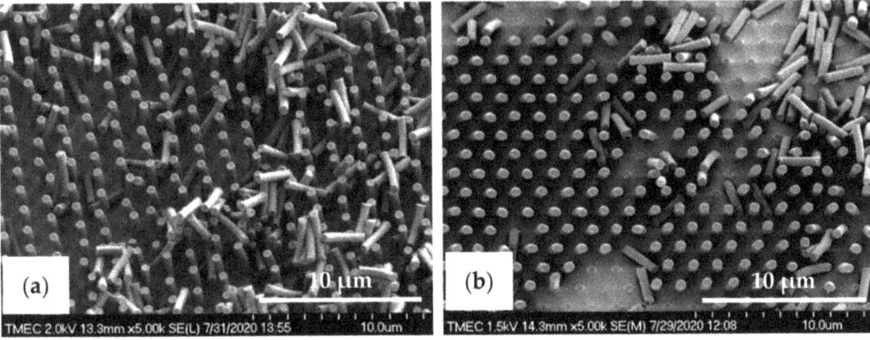

Figure 20. Top-view SEM images of the R2P NIL replicated pillar micro-structure (designed pattern size: diameter: 0.5 µm, height: 2.5 µm) after scratch test (**a**) imprint on PC foil with resin A (PIL-A) and (**b**) imprint on PC foil with resin B (PIL-B).

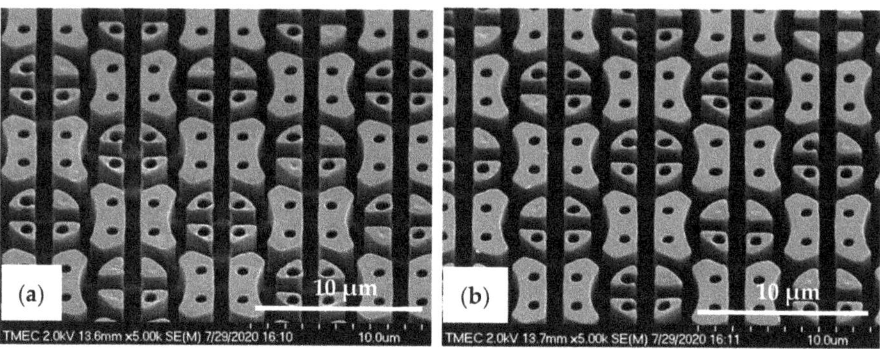

Figure 21. Tilted-view SEM images of the R2P NIL replicated C-RESS micro-structure (pattern design: size: 1.5 µm, space: 0.5 µm, height: 2.5 µm) after scratch test (**a**) imprint on PC foil with resin A (C-RESS-A) and (**b**) imprint on PC foil with resin B (C-RESS-B).

For both samples, there is no pattern collapsing, twisting or breaking off observed. This result shows that the mechanical strength of the used imprint resins is sufficiently high when used in combination with the robustly designed C-RESS micro-structure pattern in order to withstand the scratch test. After the scratch tests, the WCA and EGCA of the PIL-A, PIL-B, C-RESS-A, and C-RESS-B imprinted micro-structures were found to be reduced to (143.3 ± 2.0° and 137.3 ± 4.0°), (127.4 ± 9.3° and 109.1 ± 12.1°), (135.8 ± 5.6° and 135.5 ± 8.2°), and (138.8 ± 5.0° and 121.9 ± 9.4°), respectively. As shown in Table 4, it was also found that the S.D. of the WCA and EGCA of the PIL-A and PIL-B micro-structures significantly increased after the scratch tests. This can be explained by the variation of the surface roughness across the sample surface due to the collapsing, clumping and breaking off of the pillars. Surprisingly, the S.D. of the WCA and EGCA of the C-RESS-A and C-RESS-B patterns increased as well after the scratch test, despite the C-RESS micro-structure being robust enough to preserve its original pattern structures. Furthermore, the WCA and EGCA of the PIL-B sample were found to be significantly lower than those of the PIL-A and even the C-RESS-A and C-RESS-B imprints after the scratch tests. Additionally, the CAH values of the PIL-B sample were much higher after the scratch test than before. Therefore, it could be that the PIL-B imprinted pattern also has a higher degree of chemical degradation of the surface besides the mechanical defects, although this is difficult to conclude from this small sample set.

Table 4. WCA and EGCA and its contact angle hysteresis (CAH) of PIL and C-RESS micro-structure with resin A and resin B on PC foil after scratch tests.

Scheme	Water Droplet		Ethylene Glycol Droplet	
	WCA [°]	CAH [°]	EGCA [°]	CAH [°]
PIL-A	143.3 ± 2.0	2.2 ± 1.4	137.3 ± 4.0	1.3 ± 1.1
PIL-B	127.4 ± 9.3	10.7 ± 0.6	109.1 ± 12.1	16.3 ± 8.0
C-RESS-A	135.8 ± 5.6	7.0 ± 5.7	135.5 ± 8.2	6.2 ± 2.1
C-RESS-B	138.8 ± 5.0	10.5 ± 1.7	121.9 ± 9.4	12.3 ± 2.4

The decrease in the WCA and EGCA of the C-RESS-A and C-RESS-B patterns might be related to the chemical degradation of the surface after the scratch tests. For example, the resin surface properties, including surface charge, surface free energy (wettability), and nano-scale surface roughness could have been altered after scratching [55]. Compared to the PIL micro-structure, the replicated C-RESS patterns have a higher (mechanical) durability resulting in a smaller decrease in the WCA and EGCA after the scratch test, as shown in Figure 22.

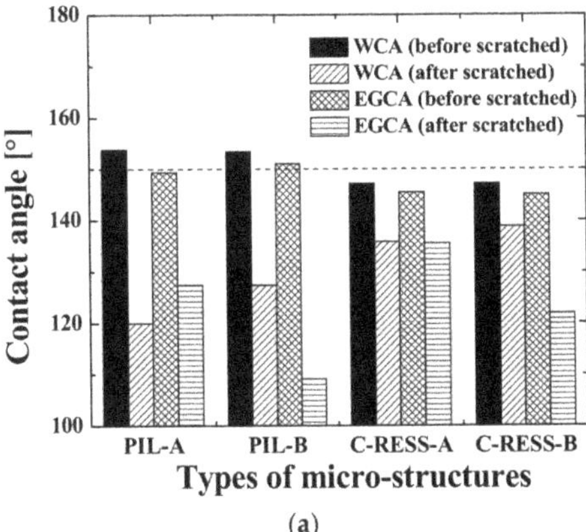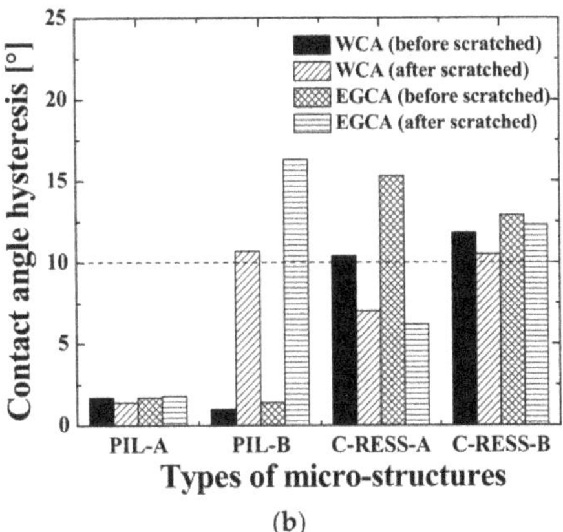

Figure 22. Comparison of the surface wetting properties of the PIL and C-RESS replicated micro-structures before and after scratch tests (**a**) water and ethylene glycol contact angle (WCA and EGCA) and (**b**) contact angle hysteresis (CAH).

There was no significant difference of the WCA and EGCA between the C-RESS-A and C-RESS-B imprints after the scratch test, even though the SFE of resin A (γ_s~15 mN/m) is higher than that of resin B (γ_s~10 mN/m). However, the CAH of the water and EG droplets on the C-RESS-B sample was found to be larger than those on the C-RESS-A sample, which could point to additional differences in the surface quality between resin A and B after the scratch test. Nevertheless, if the R2P NIL replication quality of the C-RESS micro-structure can be improved, the initial surface wettability will be better and the small decrease in the WCA and EGCA after scratch test could be acceptable for robust large-area antifouling application.

In summary, based on the above-mentioned results, the C-RESS micro-structure fabricated by R2P NIL using resin A is considered attractive for robust large-area antifouling purposes because of three main benefits. Firstly, although the high A.R. C-RESS micro-structure was challenging to imprint, the pattern fidelity was not too bad (<5% shrinkage compared to the pattern design) so that the A.R. remained close to 5.0. These imprints already result in a hydro- and oleophobicity with the WCA and EGCA of up to 144.6 ± 2.9° and 144.5 ± 2.7° before scratch test, respectively. These values could even be higher if the replication quality can be improved by reducing the amount of filled holes. Moreover, Morphotonics can scale up this pattern from the 6 inch Si wafers to square-meter sized areas, which can then be imprinted more than a thousand times using a single flexible stamp, making the R2P NIL process highly cost-effective.

Secondly, the unique surface topology of the C-RESS pattern may create an air cushion layer even when the surface did not qualify as a superhydrophobic and superoleophobic surface in the Cassie-Baxter regime (WCA and EGCA > 150°). This air cushion layer changes the solid–liquid interface of the C-RESS micro-structure to a combination of solid-liquid and air-liquid interfaces [25,56]. Therefore, the contact area between the micro-organisms and the C-RESS pattern is reduced. Consequently, the surface adhesion force of the micro-organisms decreases and the interaction strength is also reduced [57]. It was also reported that the high A.R. C-RESS pattern with hydrophobic and oleophobic properties reduced the adhesion strength of the adhesive layer, called extracellular polymeric substance (EPS), which can reduce the formation of a biofilm and subsequently suppresses the accumulation of micro-organisms on the material surface [36].

Thirdly, due to its robustness, the imprinted C-RESS pattern can maintain its pattern shape during the scratch tests and, thereby, largely its hydrophobicity and oleophobicity. It can be concluded that the robust C-RESS micro-structure fabricated with a resin having properties like resin A by using the R2P NIL process available at Morphotonics is expected to be one of the promising robust antifouling technologies for large-area medical and marine applications.

Nevertheless, the long-term mechanical durability and chemical stability of the C-RESS micro-structure and its hydro- and oleophobicity are still among the most important challenges. It has been investigated that the dewetting properties and the liquid contact angles of the (conventional) micro-structures can change over time in laboratory environments already and can severely degrade in realistic application scenarios under various working conditions (e.g., low or high alkalinity, salinity, ions, photodegradation, and high temperatures) [58,59]. This might be due to the degradation of the micro-structure patterns and the alteration of the chemical composition of the top surface, which increase the material surface energy and decrease the hydro- and oleophobic properties [58]. However, our previous report showed that the unique surface topology and durability of the PDMS-C-RESS micro-structure can maintain the level of the surface roughness and its hydrophobicity during a field test in seawater environment for 5 months [36]. In future work, such tests should also be conducted for the R2P NIL fabricated C-RESS patterns in order to investigate the performance and long-term mechanical and chemical stability in the field of the samples made by this method.

4. Conclusions

The robustness of micro- or nano-structures with a high aspect ratio (A.R.) and the scalability of the large-area patterning and mass production of it, are one of the most critical issues to produce cost-effective antifouling surfaces with both superhydrophobic and superoleophobic properties. In this research, a micro-pillar pattern of A.R. 5.0 was successfully replicated by Morphotonics' R2P NIL process on their automated Portis NIL600 system using in-house UV-curable resins with low SFE onto PC and PET foil substrates, as an alternative fabrication method for the conventional PDMS soft lithography process. The imprinted micro-pillar (PIL) pattern of A.R. 5.0 showed excellent replication quality without defects and good pattern fidelity (<6% pattern height deviation) which resulted in superhydrophobicity and oleophobicity (WCA above 150°). Furthermore, the well-designed robust circular rings with eight stripe supporters (C-RESS) pattern of A.R. 5.0 was replicated with the same process and materials. The replication quality of this challenging C-RESS imprint pattern was not perfect yet showing defects like filled holes and delaminated features due to large delamination forces. Nevertheless, the imprinted C-RESS micro-structure exhibited a good pattern fidelity (<5% pattern height deviation), already resulting in high hydro- and oleophobicity (WCA above 145°). In addition, the imprinted C-RESS pattern has a good durability during scratch tests, thereby maintaining its pattern shape and largely also its hydro- and oleophobicity. The R2P NIL results have been compared to PDMS samples fabricated by a (conventional) soft lithography process. These samples have a lower WCA due to lower A.R. and lower replication quality.

Therefore, the R2P NIL process available at Morphotonics is expected to be a promising patterning process to fabricate large-scale C-RESS micro-structure patterned substrates with antifouling properties in practical utilization for medical and marine applications.

Author Contributions: Conceptualization, N.A., M.D., P.V. and J.M.t.M.; methodology, W.S., P.P., R.M. and J.S.; Si mold fabrication and soft lithography process, P.S., O.T., N.R., K.P. and W.U.; characterizations, N.A. and M.D.; data analysis and writing—original draft preparation, N.A., M.D., N.K., W.J., P.V., J.M.t.M., review and editing. All authors have read and agreed to the published version of the manuscript.

Funding: This research was funded by the National Electronics and Computer Technology Center (NECTEC), Thailand (FleXARs project, grant no. P1951452).

Institutional Review Board Statement: Not applicable.

Informed Consent Statement: Not applicable.

Data Availability Statement: All results were analyzed by two independent observers. Data were measured at least two different locations on the sample and presented as mean value ± standard deviation (SD). All data included in this study are available upon request by contact with the corresponding author.

Acknowledgments: The authors would like to thank all TMEC staff for the silicon mold fabrication, PDMS patterning by soft lithography process and sample characterization. The authors express their gratitude to all Morphotonics B.V. staff for the large-area patterning by using Morphotonics' Portis NIL600 automated roll-to-plate nanoimprint lithography (R2P NIL) system. The authors also would like to thank KRÜSS GmbH for the surface energy measurement using the mobile surface analyzer (MSA) system.

Conflicts of Interest: The authors declare no conflict of interest.

References

1. Bixler, G.D.; Bhushan, B. Biofouling: Lessons from nature. *Philos. Trans. R. Soc. A* **2012**, *370*, 2381–2417. [CrossRef] [PubMed]
2. Lu, N.; Zhang, W.; Weng, Y.; Chen, X.; Cheng, Y.; Zhou, P. Fabrication of PDMS surfaces with micro patterns and the effect of pattern sizes on bacteria adhesion. *Food Control* **2016**, *68*, 344–351. [CrossRef]
3. Callow, J.A.; Callow, M.E. Trends in the development of environmentally friendly fouling-resistant marine coatings. *Nat. Commun.* **2011**, *2*, 1–10. [CrossRef] [PubMed]
4. Saber, A.; Mortensen, A.; Szarek, J.; Jacobsen, N.; Levin, M.; Koponen, I.; Jensen, K.; Vogel, U.; Wallin, H. Toxicity of pristine and paint-embedded TiO_2 nanomaterials. *Hum. Exp. Toxicol.* **2019**, *38*, 1–25. [CrossRef] [PubMed]
5. Ciriminna, R.; Bright, F.V.; Pagliaro, M. Ecofriendly antifouling marine coatings. *ACS Sustain. Chem. Eng.* **2015**, *3*, 559–565. [CrossRef]
6. Kim, I.-S.; Baek, M.; Choi, S.-J. Comparative cytotoxicity of Al_2O_3, CeO_2, TiO_2 and ZnO nanoparticles to human lung cells. *J. Nanosci. Nanotechnol.* **2010**, *10*, 3453–3458. [CrossRef] [PubMed]
7. Antizar-Ladislao, B. Environmental levels, toxicity and human exposure to tributyltin (TBT)-contaminated marine environment: A review. *Environ. Int.* **2008**, *34*, 292–308. [CrossRef]
8. Genzer, J.; Efimenko, K. Recent developments in superhydrophobic surfaces and their relevance to marine fouling: A review. *Biofouling* **2006**, *22*, 339–360. [CrossRef]
9. Vladkova, T.G.; Akuzov, D.T.; Klöppel, A.; Brümmer, F. Current approaches to reduction of marine biofilm formation. *J. Chem. Technol. Metall.* **2014**, *49*, 345–355.
10. Graham, M.V.; Cady, N.C. Nano and microscale topographies for the prevention of bacterial surface fouling. *Coatings* **2014**, *4*, 37–59. [CrossRef]
11. Latthe, S.S.; Gurav, A.B.; Maruti, C.S.; Vhatkar, R.S. Recent progress in preparation of superhydrophobic surfaces: A review. *J. Surf. Eng. Mater. Adv. Technol.* **2012**, *2*, 76–94. [CrossRef]
12. Latthe, S.S.; Terashima, C.; Nakata, K.; Fujishima, A. Superhydrophobic surfaces developed by mimicking hierarchical surface morphology of lotus leaf. *Molecules* **2014**, *19*, 4256–4283. [CrossRef] [PubMed]
13. Barthlott, W.; Mail, M.; Bhushan, B.; Koch, K. Plant surfaces: Structures and functions for biomimetic innovations. *Nano Micro Lett.* **2017**, *9*, 23. [CrossRef] [PubMed]
14. Bhushan, B.; Jung, Y.C. Wetting study of patterned surfaces for superhydrophobicity. *Ultramicrosc. J.* **2007**, *107*, 1033–1041. [CrossRef]
15. Ma, S.; Ye, Q.; Pei, X.; Wang, D.; Zhou, F. Antifouling on Gecko's feet inspired fibrillar surfaces: Evolving from land to marine and from Liquid repellency to algae resistance. *Adv. Mater. Interfaces* **2015**, *2*, 1500257. [CrossRef]
16. Lin, X.; Hong, J. Recent advances in robust superwettable membranes for oil–water separation. *J. Adv. Mater. Interfaces* **2019**, 1900126. [CrossRef]
17. Teisala, H.; Tuominen, M.; Kuusipalo, J. Adhesion mechanism of water droplets on hierarchically rough superhydrophobic rose petal surface. *J. Nanomater.* **2011**, *2011*, 818707. [CrossRef]
18. Bayer, I.S. Superhydrophobic coatings from ecofriendly materials and processes: A review. *Adv. Mater. Interfaces* **2020**, *7*, 2000095. [CrossRef]
19. Gao, N.; Yan, Y. Modeling superhydrophobic contact angles and wetting transition. *J. Bionic Eng.* **2009**, *6*, 335–340. [CrossRef]
20. Gogolides, E.; Ellinas, K.; Tserepi, A. Hierarchical micro and nano structured, hydrophilic, superhydrophobic and superoleophobic surfaces incorporated in microfluidics, microarrays and lab on chip microsystems. *Microelectron. Eng.* **2015**, *132*, 135–155. [CrossRef]
21. Avramescu, R.E.; Ghica, M.V.; Dinu-Pîrvu, C.; Prisada, R.; Popa, L. Superhydrophobic natural and artificial surfaces- A structural approach. *Materials* **2018**, *11*, 866. [CrossRef] [PubMed]
22. Chen, F.; Song, J.; Lu, Y.; Huang, S.; Liu, X.; Sun, J.; Carmalt, C.J.; Parkin, I.P.; Xu, W. Creating robust superamphiphobic coatings for both hard and soft materials. *J. Mater. Chem. A* **2015**, *3*, 20999–21008. [CrossRef]
23. Wenzel, R.N. Resistance of solid surfaces to wetting by water. *Ind. Eng. Chem.* **1936**, *28*, 988–994. [CrossRef]

24. Jin, J.E.Y.; Deng, Y.; Zuo, W.; Zhao, X.; Han, D.; Peng, Q.; Zhang, Z. Wetting models and working mechanisms of typical surfaces existing in nature and their application on superhydrophobic surfaces: A review. *Adv. Mater. Interfaces* **2017**, 1701052. [CrossRef]
25. Liu, M.; Wang, S.; Jiang, L. Nature-inspired superwettability systems. *Nat. Rev. Mater.* **2019**, *2*, 17036. [CrossRef]
26. Tuteja, A.; Choi, W.; McKinley, G.H.; Cohen, R.E.; Rubner, M.F. Design parameters for superhydrophobicity and superoleophobicity. *MRS Bull.* **2008**, *33*, 752–758. [CrossRef]
27. Atthi, N.; Nimittrakoolchai, O.; Supothina, S.; Kittikul, S.; Prawetpai, N.; Supadech, J.; Jeamsaksiri, W.; Pankiew, A.; Hruanun, C.; Poyai, A. An effect of silicon micro-patterning arrays on superhydrophobic surface. *J. Nanosci. Nanotechnol.* **2011**, *11*, 1–7. [CrossRef]
28. Simpson, J.T.; Hunter, S.R.; Aytug, T. Superhydrophobic materials and coatings: A review. *Rep. Prog. Phys.* **2015**, *78*, 086501. [CrossRef]
29. Kwon, B.; Kim, J.H. Importance of molds for nanoimprint lithography: Hard, soft, and hybrid molds. *J. Nanosci.* **2016**, *2016*, 1–12. [CrossRef]
30. Sabbah, A.; Youssef, A.; Damman, P. Superhydrophobic Surfaces Created by Elastic Instability of PDMS. *Appl. Sci.* **2016**, *6*, 152. [CrossRef]
31. Zhang, H.; Lamb, R.; Lewis, J. Engineering nanoscale roughness on hydrophobic surface—Preliminary assessment of fouling behavior. *Sci. Technol. Adv. Mater.* **2005**, *6*, 236–239. [CrossRef]
32. Kim, J.U.; Lee, S.; Kim, T.-I. Recent advances in unconventional lithography for challenging 3D hierarchical structures and their applications. *J. Nanomater.* **2016**, 1–17. [CrossRef]
33. Atthi, N.; Sripumkhai, W.; Pattamang, P.; Thongsook, O.; Meananeatra, R.; Saengdee, P.; Srihapat, A.; Supadech, J.; Janseng, T.; Maneesong, A.; et al. Superhydrophobic and superoleophobic properties enhancement on PDMS micro-structure using simple flame treatment method. *Microelectron. Eng.* **2020**, *230*, 111362. [CrossRef]
34. Atthi, N.; Sripumkhai, W.; Pattamang, P.; Thongsook, O.; Meananeatra, R.; Saengdee, P.; Ranron, N.; Srihapat, A.; Supadech, J.; Klunngien, N.; et al. Superhydrophobic property enhancement on guard ring micro-patterned PDMS with simple flame treatment. *Jpn. J. Appl. Phys.* **2020**, *59*, SIIJ05. [CrossRef]
35. Bae, W.-G.; Kwak, M.K.; Jeong, H.E.; Pang, C.; Jeong, H.; Suh, K.Y. Fabrication and analysis of enforced dry adhesives with core–shell micropillars. *Soft Matter.* **2013**, *9*, 1422–1427. [CrossRef]
36. Atthi, N.; Sripumkhai, W.; Pattamang, P.; Thongsook, O.; Srihapat, A.; Meananeatra, R.; Supadech, J.; Klunngien, N.; Jeamsaksiri, W. Fabrication of robust PDMS micro-structure with hydrophobic and antifouling properties. *Microelectron. Eng.* **2020**, *224*, 111255. [CrossRef]
37. Dou, Q.; Wang, C.; Cheng, C.; Han, W.; Thüne, P.C.; Ming, W. PDMS-modified polyurethane films with low water contact angle hysteresis. *Macromol. Chem. Phys.* **2006**, *207*, 2170. [CrossRef]
38. Xie, Q.; Ma, C.; Liu, C.; Ma, J.; Zhang, G. Poly(dimethylsiloxane)-based polyurethane with chemically attached antifoulants for durable marine antibiofouling. *Appl. Mater. Interfaces.* **2015**, *7*, 21030. [CrossRef]
39. Li, Y.; John, J.; Kolewe, K.W.; Schiffman, J.D.; Carter, K.R. Scaling up nature—Large area flexible biomimetic surfaces. *ACS Appl. Mater. Interfaces* **2015**, *7*, 23439–23444. [CrossRef]
40. Cui, Y.H.; Paxson, A.T.; Smyth, K.M.; Varanasi, K.K. Hierarchical polymeric textures via solvent-induced phase transformation: A single-step production of large-area superhydrophobic surfaces. *Colloids Surf. A* **2012**, *394*, 8–13. [CrossRef]
41. Xue, C.H.; Jia, S.T.; Zhang, J.; Ma, J.Z. Large-area fabrication of superhydrophobic surfaces for practical applications: An overview. *Sci. Technol. Adv. Mater.* **2010**, *11*, 033002. [CrossRef]
42. Tan, H.; Gilbertson, A.; Chou, S.Y. Roller nanoimprint lithography. *J. Vac. Sci. Technol. B Microelectron. Nanometer Struct. Process. Meas. Phenom.* **1998**, *16*, 3926–3928. [CrossRef]
43. Ahn, S.H.; Guo, L.J. Large-area roll-to-roll and roll-to-plate nanoimprint lithography: A step toward high-throughput application of continuous nanoimprinting. *ACS Nano* **2009**, *3*, 2304–2310. [CrossRef]
44. Kooy, N.; Mohamed, K.; Pin, L.T.; Guan, O.S. A review of roll-to-roll nanoimprint lithography. *Nanoscale Res. Letts.* **2014**, *9*, 1–13. [CrossRef]
45. Yong, J.; Chen, F.; Yang, Q.; Huoa, J.; Hou, X. Superoleophobic surfaces. *Chem. Soc. Rev.* **2017**, *46*, 4168–4217. [CrossRef]
46. Kern, W. The evolution of silicon wafer cleaning technology. *J. Electrochem. Soc.* **1990**, *137*, 1887–1892. [CrossRef]
47. Nanoimprint machines | Morphotonics. Available online: https://www.morphotonics.com/nanoimprint-machines/ (accessed on 1 December 2020).
48. Li, F.; Du, M.; Zheng, Z.; Song, Y.; Zheng, Q. A facile, multifunctional, transparent, and superhydrophobic coating based on a nanoscale porous structure spontaneously assembled from branched silica nanoparticles. *Adv. Mater. Interfaces* **2015**, *2*, 1500201. [CrossRef]
49. Lam, C.N.; Wu, R.; Li, D.; Hair, M.; Neumann, A. Study of the advancing and receding contact angles: Liquid sorption as a cause of contact angle hysteresis. *Adv. Colloid Interface Sci.* **2002**, *96*, 169–191. [CrossRef]
50. Owens, D.K.; Wendt, R.C. Estimation of the surface free energy of polymers. *J. Appl. Polym. Sci.* **1969**, *13*, 1741–1747. [CrossRef]
51. Xu, T.; Tao, Z.; Li, H.; Tan, X.; Li, H. Effects of deep reactive ion etching parameters on etching rate and surface morphology in extremely deep silicon etch process with high aspect ratio. *Adv. Mech. Eng.* **2017**, *9*, 1–19. [CrossRef]

52. Gosalvez, M.A.; Zhou, Y.; Zhang, Y.; Zhang, G.; Li, Y.; Xing, Y. Simulation of microloading and ARDE in DRIE. In Proceedings of the 18th International Conference on Solid-State Sensors, Actuators and Microsystems (TRANSDUCERS), Anchorage, AK, USA, 21–25 June 2015. [CrossRef]

53. Karttunen, J.; Kiihamäki, J.; Franssila, S. Loading effects in deep silicon etching. *SPIE Micromach. Microfabr. Proc. Technol. VI* **2000**, *4174*, 90–97. [CrossRef]

54. Atthi, N.; Na Ubol, P.; Sripumkhai, W.; Pattamang, P.; Meananeatra, R.; Supadech, J.; Jeamsaksiri, W.; Hruanun, C. Fabrication of PDMS with 3D microstructure for antibacterial antifouling applications. In Proceedings of the 65th JSAP Spring Meeting, Tokyo, Japan, 9 March 2018; pp. 6–57.

55. Brzozowska, A.M.; Maassen, S.; Rong, R.G.Z.; Benke, P.I.; Lim, C.S.; Marzinelli, E.M.; Jańczewski, D.; Teo, S.L.M.; Vancso, G.J. Effect of variations in micropatterns and surface modulus on marine fouling of engineering polymers. *ACS Appl. Mater. Interfaces* **2017**, *9*, 17508–17516. [CrossRef] [PubMed]

56. Magin, C.M.; Cooper, S.P.; Brennan, A.B. Non-toxic antifouling strategies. *Mater. Today* **2010**, *13*, 36–44. [CrossRef]

57. Hwang, G.B.; Page, K.; Patir, A.; Nair, S.P.; Allan, E.; Parkin, I.P. The anti-biofouling properties of superhydrophobic surfaces are short-lived. *ACS Nano* **2018**, *12*, 6050–6058. [CrossRef] [PubMed]

58. Darband, G.B.; Aliofkhazraei, M.; Khorsand, S.; Sokhanvar, S.; Kaboli, A. Science and engineering of superhydrophobic surfaces: Review of corrosion resistance, chemical and mechanical stability. *Arab. J. Chem.* **2020**, *13*, 1763–1802. [CrossRef]

59. Jing, X.; Guo, Z. Biomimetic super durable and stable surfaces with superhydrophobicity. *J. Mater. Chem. A* **2018**, *6*, 16731–16768. [CrossRef]

 *nanomaterials* 

Article

Plasmon-Assisted Direction- and Polarization-Sensitive Organic Thin-Film Detector

Michael J. Haslinger [1,2,*], Dmitry Sivun [2,3], Hannes Pöhl [2], Battulga Munkhbat [2,4], Michael Mühlberger [1], Thomas A. Klar [2], Markus C. Scharber [5] and Calin Hrelescu [2,6]

[1] PROFACTOR GmbH, Functional Surfaces and Nanostructures, 4407 Steyr-Gleink, Austria; michael.muehlberger@profactor.at
[2] Institute of Applied Physics, Johannes Kepler University, 4040 Linz, Austria; Dmitry.Sivun@fh-linz.at (D.S.); hannes.poehl@protonmail.com (H.P.); battulga@chalmers.se (B.M.); thomas.klar@jku.at (T.A.K.); HRELESCC@tcd.ie (C.H.)
[3] School of Medical Engineering and Applied Social Sciences, University of Applied Sciences Upper Austria, Garnisonstraße 21, 4020 Linz, Austria
[4] Department of Physics, Chalmers University of Technology, 41296 Göteborg, Sweden
[5] Linz Institute for Organic Solar Cells/Institute of Physical Chemistry, Johannes Kepler University, 4040 Linz, Austria; markus_clark.scharber@jku.at
[6] School of Physics and CRANN, Trinity College Dublin, Dublin, Ireland
* Correspondence: Michael.Haslinger@Profactor.at; Tel.: +43-7252-885-422

Received: 18 August 2020; Accepted: 11 September 2020; Published: 17 September 2020

Abstract: Utilizing Bragg surface plasmon polaritons (SPPs) on metal nanostructures for the use in optical devices has been intensively investigated in recent years. Here, we demonstrate the integration of nanostructured metal electrodes into an ITO-free thin film bulk heterojunction organic solar cell, by direct fabrication on a nanoimprinted substrate. The nanostructured device shows interesting optical and electrical behavior, depending on angle and polarization of incidence and the side of excitation. Remarkably, for incidence through the top electrode, a dependency on linear polarization and angle of incidence can be observed. We show that these peculiar characteristics can be attributed to the excitation of dispersive and non-dispersive Bragg SPPs on the metal–dielectric interface on the top electrode and compare it with incidence through the bottom electrode. Furthermore, the optical and electrical response can be controlled by the organic photoactive material, the nanostructures, the materials used for the electrodes and the epoxy encapsulation. Our device can be used as a detector, which generates a direct electrical readout and therefore enables the measuring of the angle of incidence of up to 60° or the linear polarization state of light, in a spectral region, which is determined by the active material. Our results could furthermore lead to novel organic Bragg SPP-based sensor for a number of applications.

Keywords: plasmons; Bragg SPPs; angle of incidence; nanoimprint lithography; grating; organic solar cell

1. Introduction

In recent years, substantial research in the field of nanophotonics has targeted the understanding and manipulation of light–matter interactions on the nanoscale. In particular, metallic nanostructures that allow the excitation of plasmons, such as localized surface plasmons (LSPs) on isolated metal nanostructures, surface plasmon polaritons (SPPs) propagating at metal–dielectric interfaces or gap plasmon polaritons (GPPs) on metal–insulator–metal interfaces and the coupling of plasmonic modes gained great attention and can be addressed and utilized [1–3]. Consequently, the generation or filtering of colors [4,5], the spatial redirection of light [6–8], extraordinary optical transmission (EOT) [9,10], as well as optical cloaking and light trapping in optoelectronic devices [11–14] have been demonstrated over the last years. Nanogratings have attracted much attention due to easy excitation of Bragg SPP plasmons by diffraction and direct coupling of the wave vector of the incoming light to the reciprocal lattice vector of a grating [1,15–17]. The coupling condition of dispersive SPPs depends on the periodicity of the grating and the angle of incidence and the polarization state of the incoming light [18]. The advantages of Bragg SPPs, such as easy excitation, the angular dependent coupling condition, polarization sensitivity, as well as the field enhancements, are widely used in surface plasmon resonance (SPR) sensors [19–22], for surface plasmon coupled light emission [23], in photo detectors [11,24] and in plasmonic solar cells [12,25–28].

Although the angle- and polarization-dependent coupling condition of Bragg plasmon excitation principally allows for the SPP plasmon-based determination of the angle of incidence (AOI) or the polarization state of light rays, nowadays, non-SPP-based devices are used to determine the AOI. Such devices often consist of bulky and complex optical systems with lenses, apertures, and sensor arrays, e.g., CCD-cameras, light field cameras, or sun sensors. Light field cameras based on the Talbot effect [29–31] simultaneously capture information on intensity and direction of the light in the far field by angle dependent intensity patterns on a photodetector generated by micro lenses or a diffraction grating [32–35]. Sun sensors often use pinholes or apertures placed in a distance in front of an active pixel array to determine the position of the sun, stars, or the location coordinates of space-crafts [36], satellites, or Mars rovers [37]. In contrast, polarization sensitive photoreceptors are used for navigation and water detection by insects in nature. For instance, desert ants (genus *Catraglyphys*) use celestial polarization patterns and a high level of processing to find direct routes into their home nest [38].

Here, we follow the examples in nature and propose the idea of an organic Bragg SPP-based device to enable angle- and polarization-sensitive detection. We investigate the combination of a flat, thin film organic solar cell with integrated periodic metallic nanostructures. Although operation of such sensors in the visible spectral range are highly desirable, their realization is challenging, and the number of studies on this specific topic is rather moderate. There are a few articles reporting on polarization and wavelength-selective photodetectors based on SPP enhanced photoconductivity in tunnel junctions [24,39,40]. For a fixed angle and a fixed wavelength, Turker et al. [41], and very recently Saito et al. [20] and Tsukagoshi et al. [42], reported on SPR sensors based on a metal grating coupler embedded onto a diode, capable of detecting changes in the refractive index of the surrounding medium without the need of an external readout setup. Senanayake et al. [11] fabricated a surface plasmon enhanced photodetector based on nanopillars, which exhibits an angle dependent photo-response due to excitation of SPPs in the infrared spectral region. In addition, studies on plasmonic solar cells with incorporated periodic metallic nanogratings need to be considered as well. However, in only a few of the studies, the angular response due to SPP coupling was correlated with the performance of plasmonic optoelectronic devices based on nanogrids [12,13,43–50]. In most studies, the devices either lack of clear plasmonic modes in the wavelength region where charge carriers are generated, or the plasmonic influence on the device performance is not significant. Moreover, most studies do not provide angle dependent measurements or discuss the discrepancies between optical and electrical measurements in detail.

In this article, we report the design and fabrication of an angle and polarization sensitive plasmonic detector based on a nanostructured thin film organic solar cell. The nanostructured semi-transparent metallic bottom and top electrodes of the cell allow for light coupling to Bragg plasmons. For both sides of incidence (bottom or top illumination), we find a unique response as a function of the AOI and the linear polarization state of light. Our findings could lead to new types of SPP-based sensors for various applications. One possible application could be the detection of the AOI similar to a sun sensor, in the visible or IR spectral region. Our device is able to detect incoming light for large AOI of at least 60° in a small and adjustable spectral range. Additionally, the design could be used as integrated organic SPR sensors, which allow direct integration into microfluidic or lab-on-a-chip devices and would enable a direct electrical readout of the signals without the need for sophisticated external readout hardware.

2. Materials and Methods

2.1. Materials

OrmoStamp was purchased from micro resist technology (Berlin, Germany). Poly(3,4-ethylenedioxythiophene):poly(styrene sulfonate) (PEDOT:PSS) Clevios P.Al 4083 was purchased from Heraeus (Hanau, Germany), diluted with water, and 0.7% Zonyl as surfactant was added. Zonyl was purchased from Sigma-Aldrich (Vienna, Austria). The poly(3-hexylthiophene-2,5-diyl) (P3HT) was purchased from Rieke metals (Lincoln, NE, USA) and the [6,6]-Phenyl C61 butyric acid methyl ester [60] PCBM was purchased from Solenne BV (Groningen, Netherlands). P3HT and PCBM were mixed in a ratio of 1:0.7, dissolved in chlorobenzene, and steered for 1 day at 70 °C. To determine the layer thicknesses, each layer was individually spin coated (4000 rpm) on a silicon substrate (as delivered) and the thickness was measured on a scratch made with an scalpel by atomic force microscope (AFM, Bruker corporation, Billerica, MA, USA). The thicknesses of the PEDOT:PSS and P3HT/PCBM layers were 15 and 86 nm, respectively.

2.2. Device Fabrication

The fabrication steps are schematically depicted in Figure 1a. Firstly, a 2D square lattice of nanostructures is imprinted on a 25×25 mm^2 glass substrate by UV-based nanoimprint lithography (UV-NIL) as described elsewhere [51]. OrmoStamp [52] was used as UV-curable nanoimprint resist. The nanostructures were selected to exhibit the $(+1, 0)$ Bragg SPP mode within the spectral region of λ = 350–650 nm, overlapping with the absorption of the P3HT/PCBM. For this purpose, a silicone NIL master with nanostructures, written with e-beam lithography and successively etched in the silicone, was used. The nanostructures have a feature size of 190 nm, in x and y, and a height of h = 275 nm, as sketched in Figure 1b. The unit cell size of the square lattice is $P_x = P_y = 360$ nm. Figure 1c shows a typical topography of the nanoimprinted substrate measured with an AFM. In Figure S1, additional AFM measurements and an electron microscopy image of the cross section through the detector are presented. The bottom electrode is a semi-transparent 25 nm thin silver (Ag) layer deposited by thermal evaporation directly onto the imprinted substrate resulting in a nanomesh that contains square holes at the bottom and square layers on top of the 275 nm high posts. Subsequently, thin PEDOT:PSS hole-conduction layer was spin coated directly on top of this nanostructured electrode and annealed for 30 min at 150 °C on a hotplate (adapted form [53]). This was followed by a thin photoactive layer (P3HT/PCBM), which was spin coated and annealed for 5 min at 80 °C on a hotplate. The nominal thickness of the PEDOT:PSS (14 nm) and P3HT/PCBM (86 nm) layer was determined by spin coating the corresponding layers on silicon substrates. On the nanostructured substrates, the spin coated active layer generates a wavy surface with a periodicity given by the underlying NIL pattern. After coating of the active layer, the topography height of around 140 nm was measured (Figure S1d). Unless mentioned otherwise, the semi-transparent top electrode consists of a 10-nm calcium (Ca) and 50-nm Ag layer deposited, in a nitrogen atmosphere in a glove box, on top of the spin-coated polymers, which gives a solid corrugated metal electrode with a topography height of about 140 nm (Figure S1). The device was encapsulated with UV-curing epoxy resin and a 130-μm-thick glass cover slide in the glove box, in order to prevent degradation. A photograph of a final device comprising 8 cells is shown in real colors in Figure 1d. The individual cells are defined by the spatial overlap of the bottom electrode and the individual top electrodes (indicated by dashed orange lines).

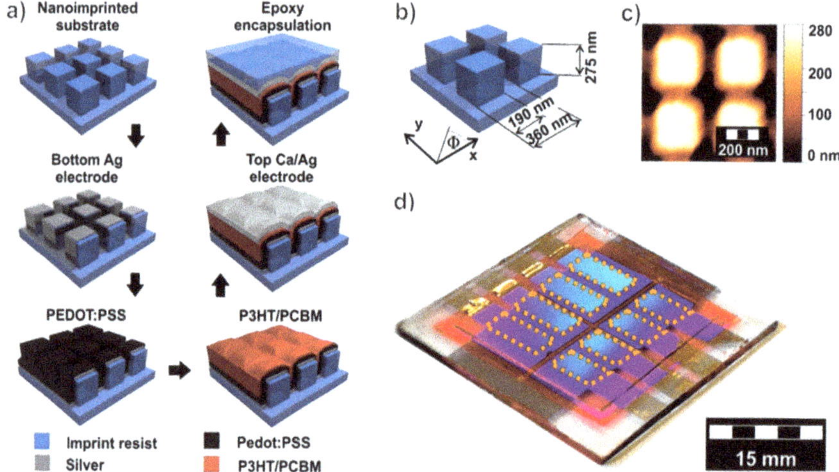

Figure 1. Manufacturing and detector details. (**a**) Fabrication of nanostructures on a glass substrate by nanoimprint lithography, followed by Ag deposition of the bottom electrode. Spin coating of PEDOT:PSS and P3HT/PCBM, metal deposition of the top electrode and encapsulation. (**b**) Schematic of a unit cell. (**c**) AFM topography image of the nanoimprinted substrate. (**d**) Photograph of the final device in true colors, comprising eight individual cells confined by the spatial overlap of the electrodes (indicated by dashed orange lines).

2.3. Optical Characterization

Angular-resolved zero-order transmission (T) and zero-order reflection (R) measurements were performed using an in-house built setup. The samples were illuminated by a broadband white light source (tungsten halogen, UV-IR) coupled to a fiber and slightly focused to a round spot with a diameter of 1 mm. T/R measurements were performed using linearly polarized light (GT 10-C Glan-Taylor Calcite Polarizer, Thorlabs, Bergkirchen, Germany) with the electric field either perpendicular (s-polarized) or parallel (p-polarized) to the plane of incidence. Polarization was controlled by turning the polarizer. All optical measurements and electrical characterization were carried out by setting the incidence plane to $\Phi = 0°$ azimuthal angle with respect to the nanostructure orientation. The measurements were performed either with incidence through the nanostructured bottom electrode (bottom electrode incidence spectra, BI) or through the semi-transparent corrugated top electrode (top electrode incidence spectra, TI). The transmitted or reflected light was collected by a lens, collimated into an optical fiber and detected with a CCD spectrometer (B&W Tek, BRX112E-V, Newak, DE, USA, range 400–1050 nm). The angle of incidence θ was varied in 2° steps for transmission measurements, which was a good tradeoff between resolution and measuring time. For each angle increment, 25 spectra were accumulated for 40-ms exposure time each. The resulting angle dependent spectra were then corrected by the response function of the setup to yield the transmittance. Reflection measurements were performed using the same setup in 5° increments. Each spectrum was accumulated 25 times with 10-ms exposure time. The resulting angle dependent spectra were again corrected by the response function of the setup to yield the reflectance.

2.4. Electrical Characterization

I–V characteristics of the devices were measured on an Oriel solar simulator using an AM1.5 spectrum. The external quantum efficiency (EQE) was determined using a setup consisting of a Xenon lamp, a monochromator, a linear-polarizer, a chopper wheel, and a lock-in amplifier (Stanford Research System SRS830, Sunnyvale, CA, USA). The samples were mounted on a manual rotation stage with the cell under investigation centered on the rotation axes illuminated with a monochromatic light spot of rectangular shape of 2×0.5 mm^2. For determining the light spectrum and the number of incoming photons, a calibrated photodetector (Hamamatsu S2281-01, Herrsching am Ammersee, Germany) was used. Angle dependent EQE measurements were performed by rotating the sample in 5° steps from 0° to 60° incident angle with 0° being the normal incidence. For each incremental step, the wavelength of the monochromatic excitation light was scanned at $\lambda = 305$–700 nm in 6-nm steps. All cells were checked for degradation effects before and after performing EQE measurements. No cell degradation was observed during the measurements and all cells showed a homogenous response over the complete active area.

2.5. Numerical Simulation

All simulations were performed in the frequency domain with the commercial finite element solver COMSOL™ (Burlington, MA, USA), version 5.2, using the Wave Optics package. The geometrical parameters are taken from scanning electron microscope (SEM, Zeiss1540 XB Oberkochen, Germany) and AFM measurements as described in Section 2.2. The refractive indices of the epoxy and the nanoimprinted structure were assumed to be purely real with a value of $n = 1.5$. The dielectric functions of silver and P3HT were taken from Johnson and Christy [54] and Ng et al. [55], respectively.

3. Results

Our proposed angle dependent detector consists of a 2D square lattice nanoimprinted on a substrate and a solution processed thin film organic solar cell on top (Figure 1). The nanostructures were selected to have Bragg SPP modes in the absorbing wavelength region of P3HT/PCBM. The top and bottom electrodes exhibit different optical properties as a consequence of their different morphologies. As stated above, the top electrode is a continuous corrugated layer with structure heights of about 140 nm, while the bottom electrode is a nanomesh that contains square holes at the bottom and square layers on top of the 275-nm high posts. Compared to an open hole array, such a blocked hole array can transmit more light, due to coupling of SPPs and LSPs, known as extraordinary optical transmission (EOT) through blocked holes [10].

To investigate the plasmonic effect of the two different electrodes on the device performance, we carried out optical and electrical characterizations of the sample as a function of the angle of incidence (AOI, θ) for both sides of illumination, top incidence (TI) and bottom incidence (BI). The optical characterization reveals the plasmon-coupling conditions, whereas the electrical characterization, especially the external quantum efficiency (EQE) measurements in the wavelength region between 350 and 650 nm, unveil the direct plasmonic influence on the charge carrier generation in P3HT/PCBM. Additionally, I–V measurements were performed on full devices in order to characterize the performance and the diode behavior.

Figure 2 shows the experimental device characterization for excitation with light incident through the top electrode (TI) in the case of p-polarization. The left panels sketch the respective experimental situation, the middle panels show the experimental results, and the right panels show the numerically simulated spectra as a function of wavelength and angle of incidence. The simulations were carried out with COMSOL Multiphysics® (cf. Section 2.5 for more details). It should be noted that the simulation results strongly depend on the properties of the materials. There is an especially big disagreement in the literature on the complex refractive index (RI) of the active material P3HT/PCBM. Figure S2 shows real and imaginary parts of P3HT/PCBM refractive index taken from three different sources [55–57]. Despite significant variations of the RI data, our simulation fits qualitatively well to the obtained experimental results in case of the RI data from Ng et al. [55], which we use for all simulations shown in the main text. The Bragg SPP coupling condition allows analytically calculating the Bragg SPP modes dispersion (cf. the Supplementary Materials). The white and black dashed lines in the panels represent the (± 1, 0) and (+2, 0) dispersive modes of the SPP at the silver/epoxy interface. They correspond well with the main measured and simulated dispersive features; however, some additional dispersive and non-dispersive features are visible in both simulations and experiments.

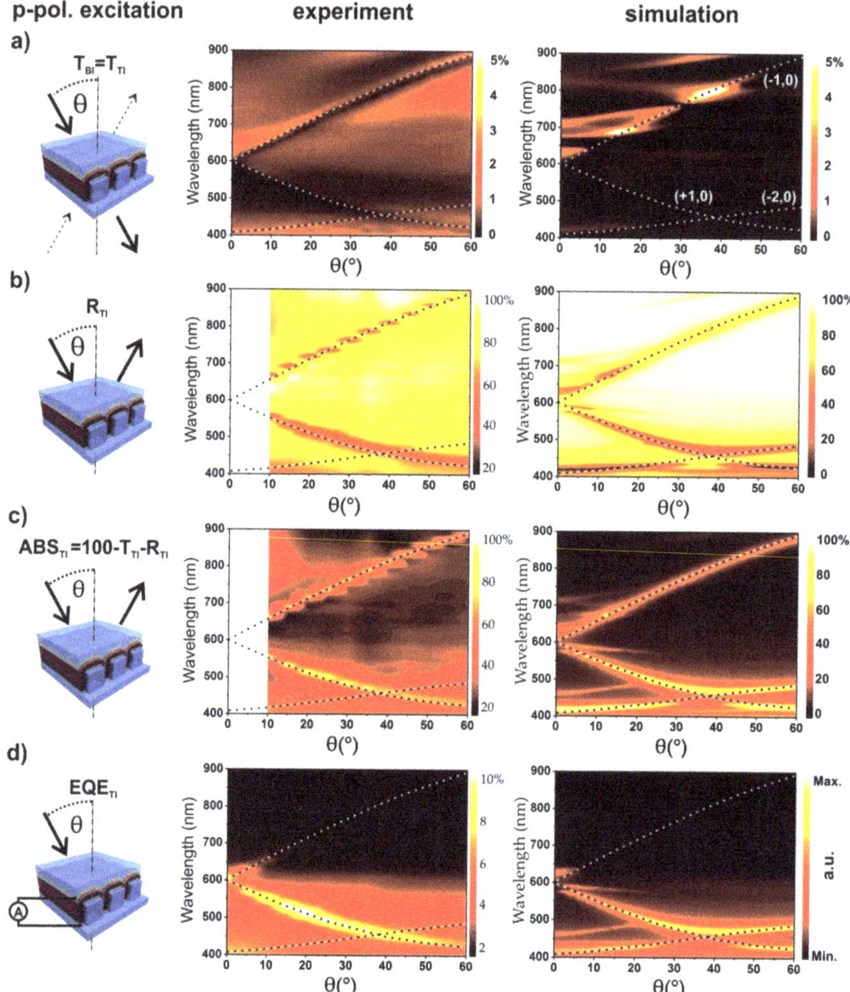

Figure 2. Schematics (left panel) for Angle-resolved device characterization for p-polarized excitation, incident on the top electrode (TI stands for top incidence). Middle panels show experimental data and right panels show corresponding simulated spectra. (**a**) Transmission (T) spectra; (**b**) reflection (R) spectra; (**c**) absorption calculated from T and R; and (**d**) external quantum efficiency (EQE) characterization. The right panel in (**d**) shows a simulation of the energy absorbed in the active layer. The abrupt decrease in EQE between 600 and 650 nm is caused by the absorption edge of the active material. The incidence plane was set to $\Phi = 0°$ azimuthal angle for all measurements. Dispersive SPP modes are clearly visible in all graphs, and calculated SPP modes are shown by dashed lines.

The total transmission through the device (Figure 2a) is below 2%, with clearly visible Bragg SPPs in the spectrum. In Figure 2b, the reflection measurements reveal that the metal top electrode acts as a mirror with a reflectivity of up to 80%. Only for light fulfilling the SPP coupling condition, the reflection drops down to 20%. Additional insights can be gained by the absorption A = 100%-T-R, as shown in Figure 2c. The experimental absorption maxima are consistent with the excitation of dispersive (±1, 0) SPP modes. Light, fulfilling the coupling conditions to the SPP, is absorbed with high efficiency within the device. Particularly, we want to address the maximum in absorption at shorter wavelengths (below 610 nm), which originates from coupling to the (+1, 0) mode. For larger AOI, anti-crossing with the (−2, 0) mode is apparent in the simulation, which is absent in the experiment. In general, energy coupled to SPP modes could be dissipated in the metal as Ohmic losses, be reemitted, or it could get absorbed in the dielectric or in the photoactive layer. Energy absorbed in the photoactive layer creates excitons, which can decay in charge carriers or recombine radiatively or non-radiatively. The external quantum efficiency (EQE) should represent a convolution of the absorption spectrum of the active layer and the plasmonic modes of the device. This can be nicely observed in Figure 2d (middle), where the measured EQE is plotted as a function of the incident wavelength and AOI. Due to the band edge and low absorption in the P3HT/PCBM above 650 nm, the (−1, 0) SPP mode is very weak in the EQE measurement, but measurable up to around 700 nm. In contrast, the (+1, 0) SPP mode is well resolved in the EQE measurement. The measured EQE at λ = 515 nm for off-mode (θ = 0°) and on-mode (θ = 20°) condition are EQE_{TI} = 3.02% and EQE_{TI} = 10.9%, respectively. This gives a relative enhancement factor of 3.6, which leads to a strong angular dependent signal for p-polarized light. At λ = 686 nm, at the band edge of P3HT/PCBM, the EQE_{TI} = 0.103% for the off-mode case and for the on-mode EQE_{TI} = 0.539%. This results in a SPP-induced enhancement of 5.2 at the band edge. One can indirectly access the EQE by simulating the absorbed energy just in the active layer (Figure 2d, right). The dispersion of the absorbed energy in the active layer is in excellent agreement with the dispersion of the experimentally measured EQE (Figure 2d, middle). Both the (+1, 0) mode and the onset of the (−1, 0) mode are well resolved. The difference is that, in the simulations more modes are prominent, the anti-crossing of the (+1, 0) mode with the (−2, 0) mode is more pronounced, and the (−2, 0) mode is clearly visible in the simulation compared to the experiments.

The optical and electrical characterizations in case of s-polarized light are shown in Figure 3. One notes that all optical and electrical characteristics at normal incidence are identical to the corresponding characteristics for p-polarization, as one expects from the symmetry of the nanostructures. The experimental and simulated transmission for the device is shown in Figure 3a. The measured transmission for s-polarization is again below 2% and shows an almost angle independent absorption edge at around 610 nm and a dispersive feature with higher transmission starting at 660 nm for θ= 15° until 790 nm for θ = 40°. The drastic change in transmission at λ = 610 nm, is caused by the change due to the absorption edge of P3HT/PCBM (see Figures S2 and S4f) and an underlying SPP mode. Reflection and absorption spectra (Figure 3b,c) show in detail the (±1, 0) SPP mode, which is only slightly dispersive in the case of s-polarization, in contrast to p-polarization. In comparison to the absorption (Figure 3c, middle), the EQE measurement (Figure 3d, middle) shows two distinct modes in the wavelength region between 550 and 650 nm. The appearance of two modes in the EQE measurement instead of one is most likely caused by a slight rotation of the sample during the measurement leading to a mode splitting.

The corresponding simulations are presented in Figure 3a–d (right). Clear Bragg SPP modes are also present in the simulation results and agree well with the measurements. However, again, the simulation results show more observable SPP modes as well as their evolution for all calculated spectra. The main (0, ±1) mode is prominent in the simulation as well as in all measurements. Two essential features of the optical properties should be highlighted here: First, for oblique incidence, a clear polarization dependent behavior is visible when comparing the EQE for p-polarization and s-polarization, (Figures 2d and 3d, respectively). Second, SPP modes excited with the s-polarization exhibit very limited dispersion in contrast to SPP modes excited with p-polarization.

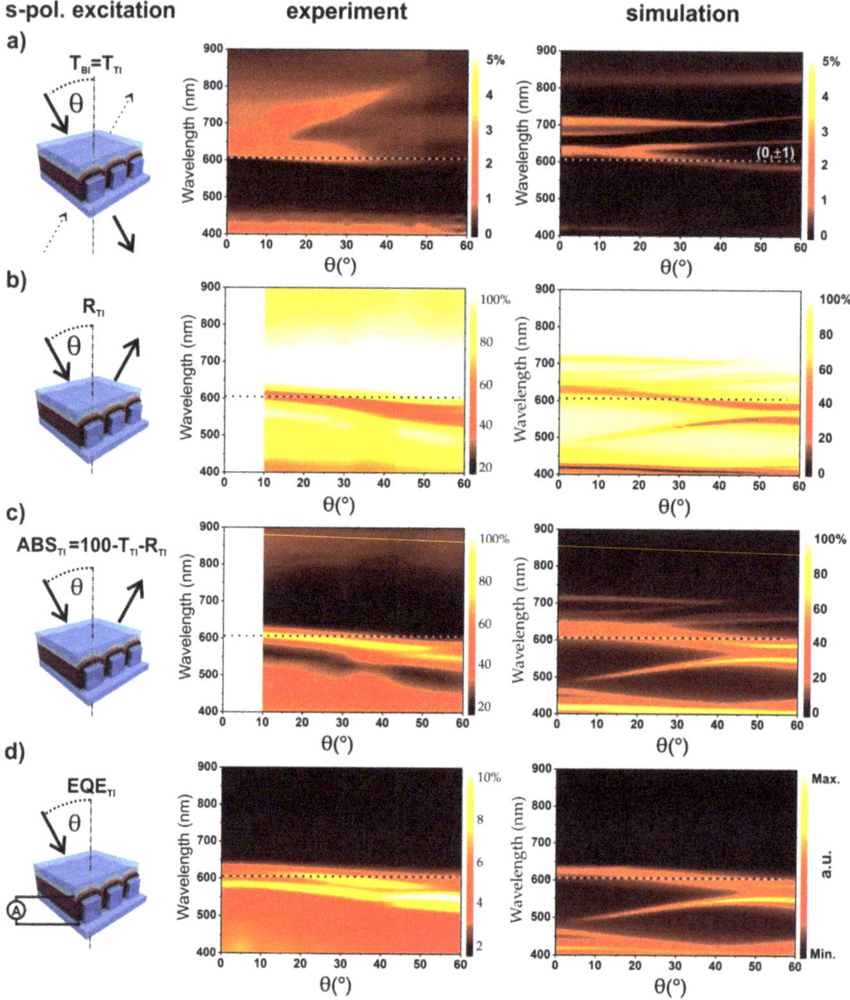

Figure 3. Schematics (left) for angle-resolved device characterization for s-polarized excitation with top incidence (TI) (middle) and corresponding simulated spectra (right). (**a**) Transmission (T) spectra; (**b**) reflection (R) spectra; (**c**) absorption measurement derived from T and R; and (**d**) external quantum efficiency (EQE) characterization. The simulation in (**d**) shows the absorbed energy in the active layer. The plateau at 600–650 nm is caused by the absorption edge of the active material overlapping with the (0, ±1) SPP mode. The incidence plane was set to $\Phi = 0°$ azimuthal angle for all measurements. SPP modes are clearly visible in all graphs, and calculated SPP modes are shown by dashed lines.

To determine the origin and location of the observed SPPs and to exclude the influences between both metal electrodes, half-cell devices were fabricated, optically characterized, and numerically simulated. A half-cell device with top electrodes only (Figure 4) was fabricated to investigate SPP modes at the metal–P3HT/PCBM and at the metal–epoxy interfaces. Additionally, devices without electrodes i.e., patterned polymer layers, were fabricated in order to get information on polymer absorption (see Figure S4).

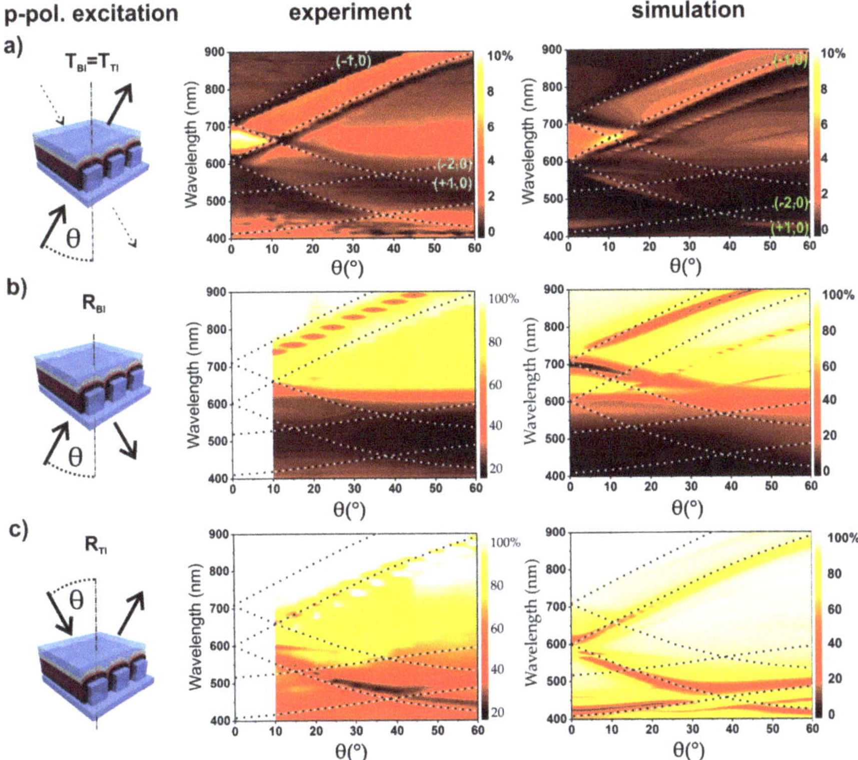

Figure 4. Schamatics (left) for side and angle resolved device characterization with p-polarized excitation (middle) and simulated spectra (right) of a half-cell device without a bottom electrode. (**a–c**) Zero-order transmission and side dependent reflection measurements. The incidence plane was set to $\Phi = 0°$ azimuthal angle for all measurements. Side dependent reflection measurement shows clear differences in SPP mode excitation. Dashed lines show the analytically calculated Bragg SPP modes at metal–epoxy (green labels in the simulation results) and metal–P3HT/PCBM interfaces (light blue labels in measurements).

Figure 4 shows the side dependent optical characterization of a half-cell device without a bottom electrode for p-polarized excitation together with analytically calculated SPP modes. The modes on the P3HT/PCBM–Ag interface are labeled in the experimental results of Figure 4a, and the modes at the metal–epoxy interface are labeled in the simulation results. The transmission measurements as well as the simulated spectra presented in Figure 4a reveal SPP modes at both metal–dielectric interfaces. The simulations disclose that the (±1, 0) SPP mode is located at the Ag–epoxy interface, starting at around $\lambda = 606$ nm for normal incidence ($\theta = 0°$), while another (±1, 0) SPP mode at the P3HT/PCBM–Ag interface starts at around $\lambda = 707$ nm for normal incidence ($\theta = 0°$). The energetic (wavelength) difference between those SPPs arises from the refractive index differences between the P3HT/PCBM, with average real part of the RI in the visible spectral range around 1.9, and the epoxy with a purely real refractive index of $n = 1.5$. The excitation of the SPPs at the respective interfaces is confirmed additionally in the reflection measurements presented in Figure 4b,c. In the case of BI excitation, the reflection spectra show primarily the SPP modes at the P3HT/PCBM–metal interface, while TI excitation predominantly couples to SPP modes at the metal–epoxy interface.

This shows that SPPs at both interfaces at the top electrode are present in the case of the half-cell device and are clearly visible in the experiments and the simulations. This is in contrast to observations from the full-cell device (Figures 2 and 3), where only the SPP modes at the metal–epoxy interface are prominent and visible, as revealed by the transmission measurements in Figure 2a. The absence of the P3HT/PCBM–metal SPP modes in the full device suggests that the bottom electrode strongly affects the Bragg SPP coupling conditions. These modes are either suppressed due to the presence of the bottom electrode or energetically shifted due to plasmonic coupling between the modes at the top and the bottom electrodes. Specifically, the part of the bottom electrode on top of the nanoimprinted pillars, which is very close to the top electrode, could affect the coupling.

Next, the influence of the bottom electrode on device characteristics is investigated by angle-resolved measurements. In Figure 5, the angle resolved device characterization and simulated spectra for p-polarized excitation are shown. The transmittance for bottom incidence is the same as for top incidence within experimental error, as presented in Figure S5. Absorption, ABS_{BI}, and EQE_{BI} are shown in Figure 5c,d, respectively. The reflection, absorption, and EQE for BI differ significantly from the corresponding results for TI. There are no clear Bragg SPP modes visible in the reflection spectra (Figure 5b). Distinct SPP modes are observed neither in the EQE nor the absorption spectra. For BI, the absorption at $\lambda = 350$–650 nm is in the range of 80%, which can be attributed to the good absorption of P3HT/PCBM in this wavelength region. The absorption and EQE spectra show a broad spectral minimum, between 500 and 700 nm, for small AOI and a broad maximum for larger AOI in the same region. For the full device, EQE lies between 16% and 29.8% over the whole spectrum and AOI range. By comparing the measured EQE_{BI} with EQE_{TI}, one notices that, in the case of TI, the energy coupled to SPP modes is generating less charge carriers. For instance, at 521 nm and 20° AOI, which represents a region with Bragg SPP coupling for TI, the external quantum efficiency for TI is $EQE_{TI(p\text{-pol})} = 10.1\%$, while for BI, the $EQE_{BI(p\text{-pol})} = 21.9\%$. Furthermore, the absorptions at 521 nm and 20° AOI for TI and BI are comparable, $ABS_{TI(p\text{-pol})} = 66.34\%$ and $ABS_{BI(p\text{-pol})} = 71.76\%$. This confirms that for TI only a part of the energy coupling to SPPs gets absorbed in the active layer, generating excitons and charge carriers. Angle resolved device characterization for BI and s-polarized excitation are presented in Figure S6.

Additionally, our results show that, for bottom incidence, our device acts as an ITO free thin film organic solar cell with no prominent angular dependency. I–V characteristics presented in the Supplementary Materials (Table S1 and Figure S7) show, for BI (TI) through the 25-nm Ag bottom electrode (10-nm Ca/50-nm Ag top electrode), a short circuit current density of $J_{sc} = 5.5$ mAcm^{-2} ($J_{sc} = 1.06$ mAcm^{-2}), a fill factor FF = 0.62 (FF = 0.56), and an open circuit voltage $V_{oc} = 0.6$ V ($V_{oc} = 0.53$ V). The difference in the device characteristics is mainly due to the different transparency of the electrodes. Furthermore, the power conversion efficiency (PCE), under AM1.5 spectrum, is $PCE_{BI} = 1.88\%$ when excited through the bottom electrode and $PCE_{TI} = 0.4\%$ when excited through the top electrode at normal incidence. The cells show a considerable PCE of up to 1.88% (see Table S1), which is an interesting demonstration for the realization of an ITO-free solar cell using metal electrodes. The efficiency is comparable with other ITO-free P3HT:PCBM organic solar cells with metal electrodes [57,58]. The presented approach might be very interesting due to its simple fabrication without any complex etching or lift-off steps and the ability of upscaling to roll-to-roll NIL processes. However, the maximum EQE of 30% is lower in comparison to optimized ITO–PEDOT:PSS–P3HT/PCBM solar cells reaching up to 60% [59]. Nevertheless, performance optimization was not within the scope of this study. The lower EQE originates from higher reflectivity of the metal bottom electrode compared to an ITO bottom electrode and different work functions.

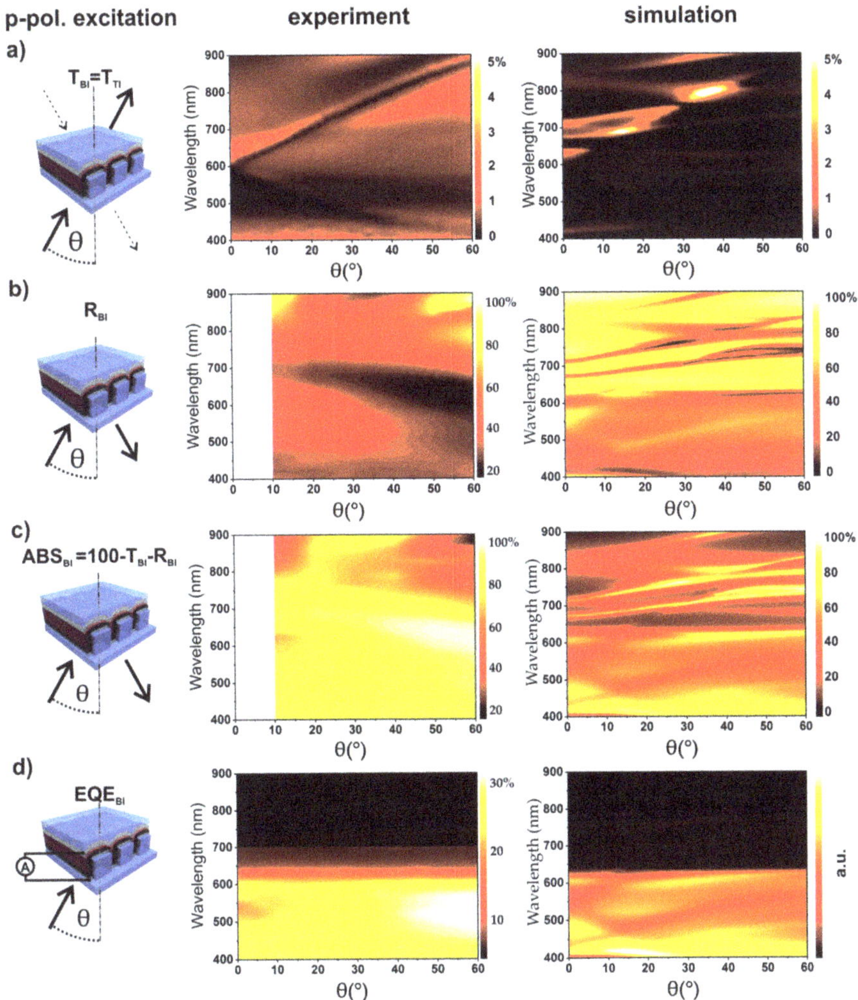

Figure 5. Schematics (left) for angle resolved device characterization with p-polarized excitation for bottom incidence (BI) (middle) and simulated spectra (right). (**a**) Transmission (T) spectra; (**b**) reflection (R) spectra; (**c**) absorption measurement derived from T and R; and (**d**) measured external quantum efficiency (EQE) characterization and the simulated absorbed energy in the active layer. The abrupt decrease in EQE between 600 and 650 nm is caused by the absorption edge of the active material. The incidence plane was set to $\Phi = 0°$ azimuthal angle for all measurements.

Since transmission spectra are independent of the side of incidence, they point out that SPPs are excited in either case of incidence. However, for BI, the influence of the SPP modes on EQE is not as prominent. This is very interesting since blocked hole arrays are known to support both SPPs and LSPs [10]. The 25-nm Ag bottom electrode exhibits a number of plasmon modes over the whole spectrum, as shown in the Figure S4 where transmission measurements of the bottom electrode without PEDOT:PSS and P3HT/PCBM layer are presented. However, measurements show that the plasmonic modes of the bottom electrode have no significant influence on the EQE of the final device. For instance, the >50% transmission at normal incidence indicates EOT since transmission through a plain solid 25-nm Ag layer gives a maximum transmission of only around 10% (e.g., λ = 500–900 nm) [60]. Our findings are in good agreement with other reports, as similar designs have been investigated before and often no significant influence due to Bragg modes in the EQE was observed for excitation through the bottom electrode [13,47]. The reason for the difference between BI and TI originates from the different shapes of the electrodes and different reflection spectra. In addition, the bottom electrodes are often made of ITO, and, if metal is used, they are usually very thin to ensure high transmission and therefore SPP modes are not as strong as for the thicker top electrodes.

EQE measurements confirm that only for top incidence, a part of the energy from the Bragg SPP modes excited at the metal–epoxy interface is absorbed in the active layer and generates charge carriers, either by direct absorption of the SPP energy in the active material or by SPP induced high transmission through the top electrode. Although the transmission measurements (Figures 2a and 5a) revealed that the same SPP modes are excited with both TI and BI, the influence on the EQE is only observed for TI. The local electric field distributions in one unit cell, corresponding to BI and TI exaction at 505 nm for various AOI, are displayed in Figure S3. The following two reasons may explain this behavior. First, for TI, the top electrode acts as a very good mirror with reflectivity reaching 80%, whereas, where coupling to SPP at the metal–epoxy interface occurs, reflection drops to 20%. The EQE measurements show that only incident light at the SPP wavelength is efficiently absorbed by the active layer and generates charge carriers. Secondly, SPPs at the top metal–epoxy interface are not excited efficiently in the case of BI, and, thus, there is no significant influence on the EQE. The reason, therefore, might be that up to 96% of the light gets absorbed in a single pass through the active layer (see Figure S4) by leading to only few percent of the incoming light reaching the metal top electrode. Indeed, this light could excite plasmons at the P3HT/PCBM–metal interface, but in contrast to TI these SPPs should lower the respected EQE. However, measurements point out that these SPPs are either not present or simply too weak to significantly contribute to a change of the EQE.

Our device architecture facilitates angle and polarization dependent response when using TI, allowing for the conceptual design of a Bragg SPP-based organic detector, sensitive to both, the AOI and the polarization state of light. The most important features enabling an angle and polarization dependent response are the corrugated semi-transparent top electrode supporting Bragg SPP modes, an active material in close contact with the nanostructured electrode providing a spatial overlap of the electromagnetic modes with the charge generation regions, and the spectral overlap of the SPP modes with the absorption of the active material which facilitates the charge generation. Light coupled to SPPs on the metal–epoxy interface and subsequently absorbed in the active material creates a photocurrent, which can be used as detector signal and is strongly linked to the coupling condition for SPPs, as can be seen in Figures 2d and 3d.

The SPP coupling condition depends on the epoxy encapsulation, the photoactive layer, the electrode materials, and the unit cell size (see Supplementary Materials). The use of epoxy encapsulation with a different refractive index could shift the coupling conditions, which allows a shift of the resonances. The active layer itself should not drastically influence SPP coupling condition, since SPPs are located at the metal–epoxy interface, but a change of the active material would change the response dramatically due to the change in the absorbing and thus charge generating spectral region. The active material P3HT/PCBM absorbs light up to 650 nm, which gives a negative angular response, since the resonance of the (+1, 0) mode blue-shifts for increasing AOI. In comparison, a detector using another active material, absorbing for example at $\lambda = 650$–900 nm, would exhibit a positive angular response due to the redshift of (−1, 0) SPP mode for increasing AOI. Our findings further show that the choice of the top electrode material and thickness strongly influences the signal strength (Figure S8). A thin top electrode shows higher intrinsic transmission but a lower signal to background ratio compared to thicker ones. The 10-nm Ca/50-nm Ag top electrode used in this work represents a good tradeoff. Another influence on the coupling condition of SPPs arises from the unit cell size of the corrugated top electrode. Smaller unit cells would blueshift the resonances, whereas larger unit cell sizes would redshift the resonances (Figure S8b). Furthermore, a rectangular unit cell or line and space pattern [39], instead of a square unit cell, would change SPP resonances for p- and s-polarization and hence could improve the polarization sensitivity.

This effect could be utilized to detect the AOI by using a narrow band excitation. The generated response strongly depends on the AOI and the polarization state of the light. It should be noted here that it is not possible to determine both azimuth and polar angle simultaneously. Therefore, for the proposed application as AOI detector, the p-polarization between emitter and detector must be guaranteed. The concept of a detector device, capable of detecting the AOI of an incoming light ray, is illustrated in Figure 6a. The proposed device would comprise a detector array of a number of individual detectors. Additionally, a linearly polarized light source with a narrow emitter wavelength is necessary. The polarization of the emitter needs to be set to p-polarization. The individual detectors would have different responses in order to be able to calculate the properties of the incoming light. The necessary different responses could either be created by using different epoxy encapsulations, different unit cell dimension of the nanostructures, or different absorber layers for the individual detectors. For example, by using nanoimprint lithography for the substrate fabrication, one could fabricate all nanostructures, in one imprint step, by imprinting a number of different detectors with different unit cell sizes. All other process steps would stay the same.

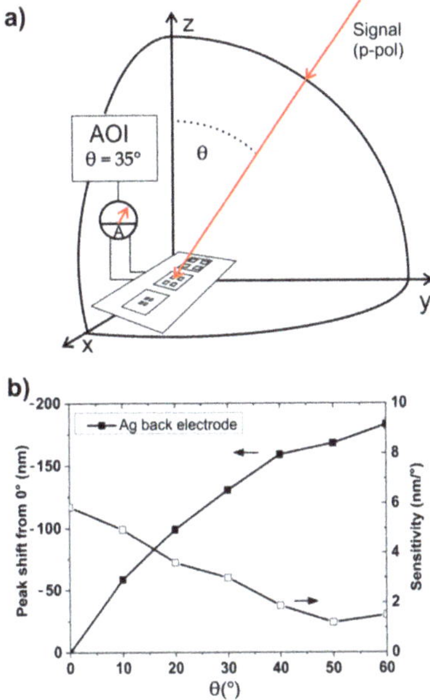

Figure 6. (**a**) Schematic illustration of the detector concept used for the detection of the AOI of incoming light. (**b**) Sensitivity and spectral shift of the resonance peak depending on the angle of incidence.

The angular sensitivity of the detector can be measured by the spectral shift of the Bragg plasmon resonance while changing the AOI by 1°. Our detector shows an angular sensitivity of up to 6 nm for p-polarized light, as can be seen in Figure 6b. The calculated SPP modes for different unit cell sizes are shown in Figure S8. We propose to take advantage of the characteristics of the $(-1, 0)$ mode, namely the dispersion over a spectral range of 180 nm. An array of 12 individual pixels (detectors) is necessary for the detection of a signal covering an AOI from 0° to 60°, by using a signal with a narrow spectral width (FWHM) of 15 nm. The benefit of our design is that such a detector can be easily fabricated directly on one substrate in one imprint process by just using a special designed master, containing an arrangement of 12 individual patterns. Additionally, at least one reference cell (with no SPP excitation) for detecting the signal strength might be necessary. A graphical solution for covering the whole AOI range is shown in Figure S9. By using a number of emitter wavelengths, a reduction of individual detectors could be possible. Even using only one detector and a wavelength sweep for the excitation, the determination of the AOI should be possible as long as the detector knows the emitted wavelength at each time. Finally, by comparing the signals of the individual detector, one could calculate the AOI of the signal.

4. Conclusions

We successfully designed and fabricated an angle of incidence (AOI) and polarization state sensitive plasmonic device. The device is based on an ITO-free thin film bulk heterojunction organic solar cell, which has two semi-transparent metal electrodes enclosing the PEDOT:PSS hole-conduction layer and photoactive layer P3HT/PCBM. The device was easily fabricated by spin coating and metal deposition on top of a nanoimprinted substrate.

A thorough analysis of the optical and electrical device characteristics, together with simulations, shows a significant influence of the side of excitation, the AOI and the polarization state of the signal on the EQE of the device. With excitation through the thin bottom electrode (BI), the device acts as an ITO-free organic solar cell with a considerable PCE of 1.88% and an EQE of up to 30% without performing any optimization. The most interesting behavior is revealed with top incidence (TI) where the device shows a clear AOI and polarization dependent response. We conclusively present that with TI the coupling of light to Bragg SPP modes on the corrugated top metal–epoxy interface occurs and that this coupling and the subsequent absorption and charge carrier generation in the device allows for an AOI and polarization dependent response. We show that SPP coupling condition can be controlled in several ways, which allows for an adaption of resonance condition, spectral range, and sensitivity. Based on our results for TI, we propose a concept of polarization sensitive Bragg SPP-based detector capable of detecting the AOI of at least up to 60°. In brief, a Bragg SPP-based AOI detector should consist of an array of individual devices showing different responses. This can be utilized by using different unit cell sizes, active materials or epoxy encapsulation materials. The individual SPP-based devices should be arranged in an array to be able to calculate the properties of the angle of the incoming light ray for p-polarized light.

In addition, our detector concept could lead to the realization of novel organic SPP-based sensors for a number of applications. For instance, our sensor concept could be used for an organic integrated SPR biosensor. By exchanging the epoxy with a thin functionalized silica protection layer and an analyte solution, one can detect changes in refractive index instead of AOI. Such a sensor would allow direct integration into microfluidic or lab-on-a-chip devices and would enable a direct electrical readout of the signal without the need of a large external setup. Furthermore, our device fabrication could be a promising strategy to demonstrate angular and polarization dependent behavior when using light emitting materials and operating the sensor as an OLED.

Supplementary Materials: The following are available online at http://www.mdpi.com/2079-4991/10/9/1866/s1, Figure S1. 3D visualizations of AFM measurements, a sketch as well as a scanning electron microscope (SEM) image of the device in different fabrication stages. Figure S2. Refractive index for P3HT:PCBM from different sources. Figure S3. Electric field distribution in one unit cell for BI and TI for various AOI. Figure S4. Side and angle-resolved characterization on different half-cell devices. Figure S5. Comparison of side dependent zero order transmission measurement for device with 25-nm bottom and 10-nm Ca/25-nm Ag top electrode. Figure S6. Angle-resolved device characterization with bottom incidence (BI) for s-polarized excitation. Table S1. Current–density–voltage characteristics of devices with 25-nm Ag front electrode and different top electrode thicknesses and materials. Figure S7. Current–voltage (J-V measurements) measurements for bottom incidence for a device with 25-nm bottom electrode and 10-nm Ca/50-nm Ag top electrode. Figure S8. Detector response (EQE) for devices with different top electrode thicknesses for normal incidence (TI) for p-polarized light. Figure S9. Necessary number of detectors depending on wavelengths to cover the complete range of AOI.

Author Contributions: Conceptualization, M.J.H., D.S., H.P., B.M., M.M., T.A.K., M.C.S. and C.H.; methodology, M.J.H., H.P., B.M., M.M., M.C.S. and C.H.; software, D.S. and C.H.; validation, M.J.H., D.S., H.P., B.M., M.M., T.A.K., M.C.S. and C.H.; investigation, M.J.H.; resources, A.K., M.C.S. and C.H.; writing—original draft preparation, M.J.H.; writing—review and editing, M.J.H., D.S., H.P., B.M., M.M., T.A.K., M.C.S. and C.H.; visualization, M.J.H. and D.S.; supervision, M.M., T.A.K., M.C.S. and C.H.; project administration, M.M., T.A.K., M.C.S. and C.H.; and funding acquisition, M.M., T.A.K., M.C.S. and C.H. All authors have read and agreed to the published version of the manuscript.

Funding: This research was funded by the Austrian Klima-und Energiefonds (SolarTrap, Grant No. 843929) and the Austrian Research and Promotion Agency (LAMPION, Grant No. 861414).

Acknowledgments: The authors would like to cordially thank Heidi Piglmayer-Brezina for the metal deposition.

Conflicts of Interest: The authors declare no conflict of interest.

References

1. Kelf, T.A.; Sugawara, Y.; Cole, R.M.; Baumberg, J.J.; Abdelsalam, M.; Cintra, S.; Mahajan, S.; Russell, A.E.; Bartlett, P.N. Localized and delocalized plasmons in metallic nanovoids. *Phys. Rev. B* **2006**, *74*, 245415. [CrossRef]

2. Törmä, P.; Barnes, W.L. Strong coupling between surface plasmon polaritons and emitters: A review. *Rep. Prog. Phys.* **2015**, *78*, 013901. [CrossRef] [PubMed]

3. Walther, R.; Fritz, S.; Müller, E.; Schneider, R.; Gerthsen, D.; Sigle, W.; Maniv, T.; Cohen, H.; Matyssek, C.; Busch, K. Coupling of Surface-Plasmon-Polariton-Hybridized Cavity Modes between Submicron Slits in a Thin Gold Film. *ACS Photonics* **2016**, *3*, 836–843. [CrossRef]

4. Gu, Y.; Zhang, L.; Yang, J.K.; Yeo, S.P.; Qiu, C.-W. Color generation via subwavelength plasmonic nanostructures. *Nanoscale* **2015**, *7*, 6409–6419. [CrossRef] [PubMed]

5. Ellenbogen, T.; Seo, K.; Crozier, K.B. Chromatic Plasmonic Polarizers for Active Visible Color Filtering and Polarimetry. *Nano Lett.* **2012**, *12*, 1026–1031. [CrossRef]

6. Ding, B.; Hrelescu, C.; Arnold, N.; Isic, G.; Klar, T.A. Spectral and Directional Reshaping of Fluorescence in Large Area Self-Assembled Plasmonic–Photonic Crystals. *Nano Lett.* **2013**, *13*, 378–386. [CrossRef]

7. Walther, B.; Helgert, C.; Rockstuhl, C.; Setzpfandt, F.; Eilenberger, F.; Lederer, F.; Tünnermann, A.; Pertsch, T.; Kley, E.-B. Spatial and spectral light shaping with metamaterials. *Adv. Mater.* **2012**, *24*, 6300–6304. [CrossRef]

8. Rodriguez, S.R.K.; Lozano, G.; Verschuuren, M.A.; Gomes, R.; Lambert, K.; De Geyter, B.; Hassinen, A.; Van Thourhout, D.; Hens, Z.; Rivas, J.G. Quantum rod emission coupled to plasmonic lattice resonances: A collective directional source of polarized light. *Appl. Phys. Lett.* **2012**, *100*, 111103. [CrossRef]

9. Ebbesen, T.W.; Lezec, H.J.; Ghaemi, H.F.; Thio, T.; Wolff, P.A. Extraordinary optical transmission through sub-wavelength hole arrays. *Nature* **1998**, *391*, 667–669. [CrossRef]

10. Li, W.-D.; Hu, J.; Chou, S.Y. Extraordinary light transmission through opaque thin metal film with subwavelength holes blocked by metal disks. *Opt. Express* **2011**, *19*, 21098–21108. [CrossRef]

11. Senanayake, P.; Hung, C.-H.; Shapiro, J.; Lin, A.; Liang, B.; Williams, B.S.; Huffaker, D.L. Surface Plasmon-Enhanced Nanopillar Photodetectors. *Nano Lett.* **2011**, *11*, 5279–5283. [CrossRef] [PubMed]

12. Nootchanat, S.; Pangdam, A.; Ishikawa, R.; Wongravee, K.; Shinbo, K.; Kato, K.; Kaneko, F.; Ekgasit, S.; Baba, A. Grating-coupled surface plasmon resonance enhanced organic photovoltaic devices induced by Blu-ray disc recordable and Blu-ray disc grating structures. *Nanoscale* **2017**, *9*, 4963–4971. [CrossRef] [PubMed]

13. Van de Groep, J.; Gupta, D.; Verschuuren, M.A.; Wienk, M.M.; Janssen, R.A.J.; Polman, A. Large-area soft-imprinted nanowire networks as light trapping transparent conductors. *Sci. Rep.* **2015**, *5*, 11414. [CrossRef] [PubMed]

14. Battaglia, C.; Hsu, C.-M.; Söderström, K.; Escarré, J.; Haug, F.-J.; Charrière, M.; Boccard, M.; Despeisse, M.; Alexander, D.T.; Cantoni, M.; et al. Light Trapping in Solar Cells: Can Periodic Beat Random? *ACS Nano* **2012**, *6*, 2790–2797. [CrossRef]

15. Kelf, T.A.; Sugawara, Y.; Baumberg, J.J.; Abdelsalam, M.; Bartlett, P.N. Plasmonic band gaps and trapped plasmons on nanostructured metal surfaces. *Phys. Rev. Lett.* **2005**, *95*, 116802. [CrossRef]

16. Hayashi, S.; Okamoto, T. Plasmonics: Visit the past to know the future. *J. Phys. D Appl. Phys.* **2012**, *45*, 433001. [CrossRef]

17. Gao, H.; Zhou, W.; Odom, T.W. Plasmonic Crystals: A Platform to Catalog Resonances from Ultraviolet to Near-Infrared Wavelengths in a Plasmonic Library. *Adv. Funct. Mater.* **2010**, *20*, 529–539. [CrossRef]

18. Ko, W.R.; Zhang, J.; Park, H.-H.; Nah, J.; Suh, J.Y.; Lee, M.H. Interfacial Mode Interactions of Surface Plasmon Polaritons on Gold Nanodome Films. *ACS Appl. Mater. Interfaces* **2016**, *8*, 20516–20521. [CrossRef]

19. Homola, J.; Yee, S.S.; Gauglitz, G. Surface plasmon resonance sensors: Review. *Sens. Actuators B Chem.* **1999**, *54*, 3–15. [CrossRef]
20. Saito, Y.; Yamamoto, Y.; Kan, T.; Tsukagoshi, T.; Noda, K.; Shimoyama, I. Electrical detection SPR sensor with grating coupled backside illumination. *Opt. Express* **2019**, *27*, 17763–17770. [CrossRef]
21. Berini, P. Surface plasmon photodetectors and their applications. *Laser Photonics Rev.* **2014**, *8*, 197–220. [CrossRef]
22. Tong, L.; Wei, H.; Zhang, S.; Xu, H. Recent Advances in Plasmonic Sensors. *Sensors* **2014**, *14*, 7959–7973. [CrossRef]
23. Toma, M.; Toma, K.; Adam, P.; Homola, J.; Knoll, W.; Dostálek, J. Surface plasmon-coupled emission on plasmonic Bragg gratings. *Opt. Express* **2012**, *20*, 14042–14053. [CrossRef] [PubMed]
24. Jestl, M.; Köck, A.; Beinstingl, W.; Gornik, E. Polarization-and wavelength-selective photodetectors. *JOSA A* **1988**, *5*, 1581–1584. [CrossRef]
25. Ou, Q.-D.; Li, Y.-Q.; Tang, J.-X. Light Manipulation in Organic Photovoltaics. *Adv. Sci.* **2016**, *3*, 1600123. [CrossRef]
26. Mandal, P.; Sharma, S. Progress in plasmonic solar cell efficiency improvement: A status review. *Renew. Sustain. Energy Rev.* **2016**, *65*, 537–552. [CrossRef]
27. Weickert, J.; Dunbar, R.B.; Hesse, H.C.; Wiedemann, W.; Schmidt-Mende, L. Nanostructured organic and hybrid solar cells. *Adv. Mater.* **2011**, *23*, 1810–1828. [CrossRef]
28. Zhou, K.; Guo, Z.; Liu, S.; Lee, J.-H. Current Approach in Surface Plasmons for Thin Film and Wire Array Solar Cell Applications. *Materials* **2015**, *8*, 4565–4581. [CrossRef] [PubMed]
29. Wang, A.; Gill, P.; Molnar, A. Light field image sensors based on the Talbot effect. *Appl. Opt.* **2009**, *48*, 5897–5905. [CrossRef]
30. Dennis, M.R.; Zheludev, N.I.; de Abajo, F.J.G. The plasmon Talbot effect. *Opt. Express* **2007**, *15*, 9692–9700. [CrossRef]
31. Zhang, W.; Zhao, C.; Wang, J.; Zhang, J. An experimental study of the plasmonic Talbot effect. *Opt. Express* **2009**, *17*, 19757–19762. [CrossRef] [PubMed]
32. Wang, A.; Gill, P.; Molnar, A. Angle sensitive pixels in CMOS for lensless 3D imaging. In Proceedings of the 2009 IEEE Custom Integrated Circuits Conference, Rome, Italy, 13–16 September 2009; pp. 371–374. [CrossRef]
33. Wang, A.; Gill, P.R.; Molnar, A. Fluorescent imaging and localization with angle sensitive pixel arrays in standard CMOS. In Proceedings of the 2010 IEEE Sensors, Kona, HI, USA, 1–4 November 2010; pp. 1706–1709. [CrossRef]
34. Johnson, B.; Peace, S.T.; Wang, A.; Cleland, T.A.; Molnar, A. A 768-channel CMOS microelectrode array with angle sensitive pixels for neuronal recording. *IEEE Sens. J.* **2013**, *13*, 3211–3218. [CrossRef]
35. Lee, C.; Johnson, B.; Molnar, A. Angle sensitive single photon avalanche diode. *Appl. Phys. Lett.* **2015**, *106*, 231105. [CrossRef]
36. Rufino, G.; Grassi, M. Multi-Aperture CMOS Sun Sensor for Microsatellite Attitude Determination. *Sensors* **2009**, *9*, 4503–4524. [CrossRef]
37. Lee, C.; Bae, S.Y.; Mobasser, S.; Manohara, H. A novel silicon nanotips antireflection surface for the micro sun sensor. *Nano Lett.* **2005**, *5*, 2438–2442. [CrossRef]
38. Marshall, J.; Cronin, T.W. Polarisation vision. *Curr. Biol.* **2011**, *21*, R101–R105. [CrossRef]
39. Jestl, M.; Maran, I.; Köck, A.; Beinstingl, W.; Gornik, E. Polarization-sensitive surface plasmon Schottky detectors. *Opt. Lett.* **1989**, *14*, 719–721. [CrossRef] [PubMed]
40. Berthold, K.; Höpfel, R.A.; Gornik, E. Surface plasmon polariton enhanced photoconductivity of tunnel junctions in the visible. *Appl. Phys. Lett.* **1985**, *46*, 626–628. [CrossRef]
41. Turker, B.; Guner, H.; Ayas, S.; Ekiz, O.O.; Acar, H.; Gulera, M.O. Grating coupler integrated photodiodes for plasmon resonance based sensing. *Lab. Chip.* **2011**, *11*, 282–287. [CrossRef]
42. Tsukagoshi, T.; Kuroda, Y.; Noda, K.; Binh-Khiem, N.; Kan, T.; Shimoyama, I. Compact Surface Plasmon Resonance System with Au/Si Schottky Barrier. *Sensors* **2018**, *18*, 399. [CrossRef]
43. You, J.; Li, X.; Xie, F.-X.; Sha, W.E.; Kwong, J.H.W.; Li, G.; Choy, W.C.H.; Yang, Y. Surface Plasmon and Scattering-Enhanced Low-Bandgap Polymer Solar Cell by a Metal Grating Back Electrode. *Adv. Energy Mater.* **2012**, *2*, 1203–1207. [CrossRef]

44. Munkhbat, B.; Pöhl, H.; Denk, P.; Klar, T.A.; Scharber, M.C.; Hrelescu, C. Performance Boost of Organic Light-Emitting Diodes with Plasmonic Nanostars. *Adv. Opt. Mater.* **2016**, *4*, 772–781. [CrossRef]
45. Liu, Y.; Kirsch, C.; Gadisa, A.; Aryal, M.; Mitran, S.; Samulski, E.T.; Lopez, R. Effects of nano-patterned versus simple flat active layers in upright organic photovoltaic devices. *J. Phys. D Appl. Phys.* **2012**, *46*, 024008. [CrossRef]
46. Richardson, B.J.; Zhu, L.; Yu, Q. Design and development of plasmonic nanostructured electrodes for ITO-free organic photovoltaic cells on rigid and highly flexible substrates. *Nanotechnology* **2017**, *28*, 165401. [CrossRef]
47. Ko, O.-H.; Tumbleston, J.R.; Zhang, L.; Williams, S.; DeSimone, J.M.; Lopez, R.; Samulski, E.T. Photonic crystal geometry for organic solar cells. *Nano Lett.* **2009**, *9*, 2742–2746. [CrossRef]
48. Hara, K.; Lertvachirapaiboon, C.; Ishikawa, R.; Ohdaira, Y.; Shinbo, K.; Kato, K.; Kaneko, F.; Baba, A. Inverted organic solar cells enhanced by grating-coupled surface plasmons and waveguide modes. *Phys. Chem. Chem. Phys.* **2017**, *19*, 2791–2796. [CrossRef] [PubMed]
49. Na, S.-I.; Kim, S.-S.; Jo, J.; Oh, S.-H.; Kim, J.; Kim, D.-Y. Efficient Polymer Solar Cells with Surface Relief Gratings Fabricated by Simple Soft Lithography. *Adv. Funct. Mater* **2008**, *18*, 3956–3963. [CrossRef]
50. Forberich, K.; Dennler, G.; Scharber, M.C.; Hingerl, K.; Fromherz, T.; Brabec, C.J. Performance improvement of organic solar cells with moth eye anti-reflection coating. *Thin Solid Films* **2008**, *516*, 7167–7170. [CrossRef]
51. Mühlberger, M.; Bergmair, I.; Klukowska, A.; Kolander, A.; Leichtfried, H.; Platzgummer, E.; Loeschner, H.; Ebm, C.; Grützner, G.; Schöftner, R. UV-NIL with working stamps made from Ormostamp. *Microelectron. Eng.* **2009**, *86*, 691–693. [CrossRef]
52. Micro Resist Technology GmbH. Available online: http://www.microresist.de/en (accessed on 10 November 2017).
53. Adam, G.; Munkhbat, B.; Denk, P.; Ulbricht, C.; Hrelescu, C.; Scharber, M.C. Different Device Architectures for Bulk-Heterojunction Solar Cells. *Front. Mater.* **2016**, *3*. [CrossRef]
54. Johnson, P.B.; Christy, R.W. Optical Constants of the Noble Metals. *Phys. Rev. B* **1972**, *6*, 4370–4379. [CrossRef]
55. Wong, W.-Y.; Li, C.H.; Fung, M.K.; Djurišić, A.B.; Zapien, J.A.; Chan, W.K.; Cheung, K.Y.; Wong, W.-Y. Accurate determination of the index of refraction of polymer blend films by spectroscopic ellipsometry. *J. Phys. Chem. C* **2010**, *114*, 15094–15101.
56. Jaglarz, J.; Małek, A.; Sanetra, J. Thermal Dependence of Optical Parameters of Thin Polythiophene Films Blended with PCBM. *Polymers* **2018**, *10*, 454. [CrossRef]
57. Stelling, C.; Singh, C.R.; Karg, M.; König, T.A.F.; Thelakkat, M.; Retsch, M. Plasmonic nanomeshes: Their ambivalent role as transparent electrodes in organic solar cells. *Sci. Rep.* **2017**, *7*, 42530. [CrossRef] [PubMed]
58. Kang, M.-G.; Kim, M.-S.; Kim, J.; Guo, L.J. Organic Solar Cells Using Nanoimprinted Transparent Metal Electrodes. *Adv. Mater.* **2008**, *20*, 4408–4413. [CrossRef]
59. Dennler, G.; Forberich, K.; Scharber, M.C.; Brabec, C.J.; Tomiš, I.; Hingerl, K.; Fromherz, T. Angle dependence of external and internal quantum efficiencies in bulk-heterojunction organic solar cells. *J. Appl. Phys.* **2007**, *102*, 054516. [CrossRef]
60. McPeak, K.M.; Jayanti, S.V.; Kress, S.J.P.; Meyer, S.; Iotti, S.; Rossinelli, A.A.; Norris, D.J. Plasmonic Films Can Easily Be Better: Rules and Recipes. *ACS Photonics* **2015**, *2*, 326–333. [CrossRef] [PubMed]

MDPI

St. Alban-Anlage 66

4052 Basel

Switzerland

Tel. +41 61 683 77 34

Fax +41 61 302 89 18

www.mdpi.com

Nanomaterials Editorial Office

E-mail: nanomaterials@mdpi.com

www.mdpi.com/journal/nanomaterials